A HANDBOOK FOR TEACHERS IN DEVELOPING COUNTRIES

WILLIAM WANJALA TOILI
Masinde Muliro University of Science and Technology

EMMANUEL WAMBOKA TOILI
St Paul's University

MARY ESTHER MUYOKA TOILI
Vrije Universiteit Brussels

Copyright © 2019 Toili Books

Cover Design: Freepik

All rights reserved.

ISBN: 9781093318555

CONTENTS

PREFACE i
FOREWORD ii
ABOUT THE AUTHORS vii
CHAPTER ONE 1
SCOPE OF BIOLOGY 1
    Introduction 2
    Definition of Biology 2
    Characteristics of Living Things 3
        Living Things have Cellular Organisation 4
        Living Things undergo Metabolism 5
        Living Things Respond to their Surroundings 5
        Living Things Move 6
        Living Things Adapt and Evolve 6
        Living Things change during their Lives (or Grow and Develop) 7
        Living Things Make More Living Things (Reproduce) 7
        Living Things Inherit Characteristics of Life from Parents 8
    The Scope of Biology: Key Themes and Principles of Biology 9
        Biologists Study Diversity of Life 9
        Biologists Study How Structure Determines Function 10
        Biologists Study How Surface Area to Volume Ratio influences Function 10
        Biologists Study Patterns of Change 10
        Biologists Study Interactions of the Environment 11
        Biologists Study Patterns of Perpetuation of Life and Development 11
        Biologists Study Unity of Life 12
        Biology Studies about Application of Technology to Biological Systems (Biotechnology) 13
    Conclusion 13
    References 14
CHAPTER TWO **15**
BIOLOGY AS A SCIENCE **15**
    Introduction 16
    The Scientific Process: The Unifying Theme of all Sciences 16
        Stages in Scientific Method 17
        Analysis and Interpretation of Data 20
        Dissemination of Results 22
        Limitations of Science and the Scientific Method 22
        Examples of How Biologists Work using the Scientific Method 23
    Fundamental Characteristics of Science 25
        Science Studies about the Real World 26
        Scientific Knowledge is Empirical 26

    Scientific Knowledge is Tentative 26
    Science uses the Scientific Method 27
    Scientific Knowledge is a Product of Logical Reasoning....................27
    Scientific Ideas are Quantifiable 29
    Science is Parsimonious 29
    Science is Honest 30
  Why Biology is considered a Science..30
    Discussion Questions..31
  Conclusion 31
  References..32

CHAPTER THREE 33
NATURE AND GOALS OF BIOLOGY EDUCATION IN SECONDARY SCHOOLS.. 33

  Introduction..34
  Nature and Meaning of Biology Education...34
  Why Teach Biology?..35
    Biology helps us to Improve Quality of Health......................................35
    Biology helps us Produce more Food 35
    Biology helps us Conserve our Natural Resources................................35
    Biology helps us to be Better Citizens...36
    Biology helps to Prepare for Careers...36
    Biology helps us to enjoy our Leisure Time 36
  Outcomes of Biology Education..37
    Scientific Knowledge...37
    Scientific Facts...37
    Scientific Attitudes...41
    Processes of Science 42
  Conclusion 49
  References..49

CHAPTER FOUR 51
THE SECONDARY SCHOOL BIOLOGY CURRICULUM............51

  Introduction..52
  The Goals of Teaching Biology in Secondary Schools................................52
  Objectives of Teaching Biology in Secondary Schools...............................52
  The Content of the Biology Curriculum in Secondary Schools 55
  The Organisation of Secondary School Biology Curriculum 56
  Development and Progress of Secondary School Biology Curriculum ..57
    Experimental Phase 57
    Adoption Phase..59
    Adaptation Phase 61
    Ownership or Indigenisation Phase..62

Factors influencing Secondary School Biology Curriculum Changes 63
    External Pressures 63
    Internal Pressures 64
Conclusion 64
Reference 64

CHAPTER FIVE 65
LEARNING BIOLOGY: THE THEORETICAL FOUNDATIONS ..
.... 65
Introduction 66
The Concept of Learning 66
How Learning Occurs: The Learning Theories 66
Some Learning Theories Relevant to Learning of Biology 67
    Behavioural Learning Theory 67
    Cognitive Learning Theory 71
    Constructivism 79
    Implications of Constructivism for Teaching Biology 83
Overall Implications of Learning Theories for Biology Teaching 84
    Design of Learning Activities 84
    Creation of Motivation 85
    Encouraging Schematic Learning 85
    Use of Reinforcement 86
    Encouraging Metacognition 86
    Learning Styles 87
    Cognitive Demand 88
    Planning 89
Conclusion 89
References 89

CHAPTER SIX 91
LEARNING IN BIOLOGY 91
Introduction 92
Approaches to Learning in Biology 92
    Surface Learning 92
    Deep Learning 94
Teaching and Learning in Biology 95
Effective Teaching: Constructive Alignment and creation of a Learning Community 96
    Barriers to Constructive Alignment 99
Concept Learning in Biology 99
    Types of Concepts 99
    Concept Learning 101
Conclusion 103
References 103

CHAPTER SEVEN 104

INSTRUCTIONAL PHILOSOPHY IN BIOLOGY **104**
   Introduction 105
   Instructional Philosophy 105
      The Key Features of Constructivism 107
   Effective Constructivist Teaching of Biology 108
   A Personal Instructional Philosophy in Biology 114
      Discussion 114
   Conclusion 115
   References 115
CHAPTER EIGHT **116**
PREPARING FOR INSTRUCTION IN BIOLOGY **116**
   Introduction 117
   Setting up the Learning Situation 117
   The Unit Plan 117
      Advantages of Unit Planning 118
      Preparations before Making a Unit Plan 118
      Characteristics of a Good Unit 119
      Steps in Unit Planning 120
      Example of Unit Plan 121
   Schemes of Work 123
      Definition of Scheme of Work 123
      The format for Schemes of Work 125
      A Sample Scheme of Work 126
   Conclusion 130
   References 131
CHAPTER NINE **132**
LESSON PLANNING IN BIOLOGY **132**
   Introduction 133
   Definition of Lesson Plan 133
   Things to be done before Lesson Planning 133
   Advantages of Planning a Lesson 134
   Characteristics of a Good Lesson Plan 135
   Procedures of Lesson Planning in Biology 135
      Stating Instructional Objectives in Biology 136
      Designing the Lesson Plan for Teaching Biology 138
   Lesson Plan Format 139
      Title/ Administrative Details 140
      Description of the Components of the Biology Lesson Plan 141
      The functions of closure are to: 143
   A Specimen Lesson Plan in Biology 147
   Conclusion 150
   References 150

CHAPTER TEN 151
METHODS, TECHNIQUES AND STRATEGIES OF TEACHING BIOLOGY 151

Introduction 152
Teaching and Learning Activities 152
Meaning of Teaching Method, Technique and Strategy 153
Factors Influencing Choice of Teaching Strategies in Biology 154
    The teacher's mastery of content, experience, personality and communication skills 154
    The size of the class 154
    Nature of content to be taught 155
    Instructional Objectives 155
    Availability of time and material resources 155
Transmission Method Techniques 156
    Discussion 156
    Homework or Assignment 160
    Demonstration 162
    Question and Answer 163
    Lecture 165
Problem Solving Method: Some selected Techniques 166
    Practical Work or Class Experiments in Biology 168
    The 5E Instructional Model in Biology 172
    Problem Based Learning in Biology 176
Project Based Learning in Biology 182
    Cooperative Learning 186
    Collaborative Learning 188
Conclusion 188
References 188

CHAPTER ELEVEN 190
INSTRUCTIONAL RESOURCES IN BIOLOGY 190

Introduction 191
Definition of Instructional Resources 191
    Types of Instructional Resources 191
    Advantages of Using Resources to Teach Biology 192
Minimum Biology Equipment and Chemicals for a Biology Room or Laboratory 193
School Equipment Production Unit (SEPU) Biology Kit 195
    Advantages of Using SEPU Biology Kits 196
Improvisation of Resources in Biology 196
Preparation of Solutions 197
Preservation of Specimens 198
    Preservation of plants 198

    Preservation of animals 198
    Preparation of Skeleton Specimens 199
    The Live Specimens in the School 200
        The School Aquarium 200
        The School Egg Incubator 200
        The School Garden 201
    Biology Text Books 201
    Teaching with Information and Communication Technology (ICT) 202
        Affordances 203
        Scaffolding 204
    Conclusion 207
    References 208

CHAPTER TWELVE 209
TEACHING BIOLOGY: APPRAISING PRACTICE 209

    Introduction 210
    Teaching Skills 210
        Set Induction 211
        Closure 211
        Organisation 212
        Planning 212
        Classroom Management 213
        Communication or Instruction 213
        Monitoring 213
        Evaluation 213
        Personal Practical Knowledge or Professional Responsibilities 213
    Self - Appraisal in Biology Instruction 214
        Quality of Performance in Biology 214
        Characteristics of Teacher Talk 216
    The 21st Century Pedagogy 217
    Transformative Learning 220
    Appraising Lessons in Biology 221
        1)   Preparation for the lesson 222
        2)   Lesson Performance 222
        3)   The personality of the Teacher 223
    Conclusion 224
    References 224

CHAPTER THIRTEEN 226
THE SCHOOL BIOLOGY LABORATORY 226

    Introduction 227
    The Design and Key Features of the School Biology Laboratory 227
    Planning for use of School Biology Laboratory 230
        Time Table and Requisitions 230

Role of Laboratory Technician in the Biology Laboratory 230
Role of the Biology Teacher in the Laboratory 231
The School Biology Laboratory Management......................... 231
Safety in the Biology Laboratory 232
   First Aid.. 232
   Common Injuries in the Biology Laboratory 232
School Biology Laboratory Rules and Regulations 234
Conclusion 235
References 236

CHAPTER FOURTEEN 237
EVALUATION OF LEARNING IN BIOLOGY 237
Introduction 238
Definition of Evaluation and Assessment 238
Types of Evaluation... 239
Purpose of Assessment in Biology... 239
Assessment Techniques in Biology.. 240
   Oral/Clinical Interviews 240
   Written Tests/Examinations.. 241
Preparing a Classroom Test in Biology 245
   The Steps 245
   Preparing Marking Scheme 247
Assessing Practical Skills... 248
   Assessment of Manipulative Skills 248
   Assessment of Science Process Skills 250
Conclusion 251
References 252

CHAPTER FIFTEEN 253
HINTS ON TEACHING SOME DIFFICULT CONCEPTS IN BIOLOGY CURRICULUM 253
Introduction 254
Some Difficult Concepts in the Biology Curriculum in Kenya............ 254
Highlights of the Key Teaching Approaches of the Difficult Biology Concepts... 256
Conclusion 269
References 270
REVISION QUESTIONS 271

PREFACE

This is the first edition of the book that focuses on equipping the pre-service and the practising teachers of biology with the current knowledge and skills in biology education. The book is a response to the demand for such a book by practising teachers, teacher trainees and trainers in secondary school biology education.

The book targets students training to become biology teachers at the Diploma, undergraduate and postgraduate levels. The book will also be a useful resource material for practising teachers of biology in secondary schools and quality assurance officers and teacher trainers in universities and colleges.

The book is based on the premise that potential teachers of biology are fairly well grounded in the various courses in botany and zoology, which provide more advanced biological knowledge than what is prescribed in the syllabi for secondary schools. The teachers are therefore required to adopt the advanced knowledge to suit the students in secondary schools. This requires the teachers to use professional approach-es that facilitate the learning of biology notwithstanding the advanced biological knowledge on the ground. This book is about how we can help learners understand and appreciate the science of life.

The book is presented in a simple, clear and Standard English language augmented with diagrammatic illustrations, pictures and tables that are intended to motivate the reader. The book has also several tasks and exercises to get the readers to reflect on what they read and to further extend their knowledge. In addition, the book provides a summary of the information at the end of each chapter to help the reader recapitulate the content of the chapter.

FOREWORD

Science, technology and innovation are the key components in the development agenda of any country. In Kenya, this is given prominence both in the medium and long-term planning as clearly articulated in Vision 2030. The fact that a good background in science and mathematics is key to entry into science-related careers like medicine, engineering, technology and agriculture that directly contribute to economic development of the country cannot be gainsaid. At the secondary school level, the science curricula are in part geared to this end.

The performance of any given curriculum in Kenya, and in many developing countries, is primarily measured by summative evaluation of students in external examinations administered to them at the end of their schooling. The secondary school students sit the Kenya Certificate of Secondary Education (KCSE) examination after four years of schooling. Every year students' performance in science (including biology) and mathematics is usually poorer than that in the social sciences and languages. Kenya's educational system may not be adequately achieving its aim of attaining self-reliance in manpower development.

Whereas many reasons may contribute to students' poor performance in science and mathematics, the major reason is certainly the passive methods of teaching that engage students in surface learning rather than deep learning. The methods don't encourage students to engage in the transformation of ideas to solve the problems encountered. The students end up cramming facts without a real understanding of what they learn. The result of this is that they are unable to apply knowledge to various new situations. They are unable to experience transformation in their lives. The use of interactive teaching techniques that a majority of teachers are unfamiliar with would certainly help remedy the situation. Such techniques are innovative in nature and their use requires teachers to be trained adequately. Lack of adequate training of teachers in science and mathematics is in part responsible for this. Clearly, there is a need to improve the quality of science education through improved training of teachers (pre-service and in-service) and provision of quality teaching and learning materials.

The universities and other teachers training colleges are key to creating ideas that can improve the quality of biology education. Improved teaching skills through innovative training programmes are a key component in this endeavour. The task of the training programmes is to empower the teachers of biology to be able to devise the simplest and most effective manner of helping the students to develop problem-solving and critical thinking skills. This would help learners to deal with a wide range of problems including mental, spiritual, metacognitive, social, economic questions as well as analysis of policies, regulations and institutional requirements thereby avoiding social problems they encounter in their daily lives.

The teaching of biology in secondary schools is certainly challenging and demanding. This places new demands on the biology teachers working in this setting. A professional revolution has taken place in the last two decades with regard to pedagogy within the wide framework of Education for All (EFA). Teachers in secondary schools in this dispensation are expected to develop a wide range of pedagogical skills to be able to meet the expectations and objectives of teaching biology in schools. The demands of biology education in secondary schools today calls for teachers who are skilled and grounded in the content of biological knowledge, values and teaching strategies that facilitate the creation of an enabling teaching and learning environment.

In this book, a selected battery of pedagogic topics aimed at increasing the knowledge and awareness of pedagogic strategies, pedagogic skills and expertise, professional attitudes as well as relating theory to practice in the teaching of biology are outlined and adequately analysed. In turn, it is our hope the pedagogic skills learned will help the biology teachers to carry out the professional responsibility amidst the huge demands they face.

The book offers an increasingly integrated conceptualization of the ingredients of a successful biology teacher in terms of the required competencies including their expanded roles intended to transform the learners into complete beings. The four pillars of education advocated by UNESCO are at the core of the teaching and learning of biology, namely, learning to know, learning to do, learning to live together and learning to be. The biology teacher ultimately aims at the creation of learning communities in which knowledge is actively co-constructed, and in which the focus of learning is sometimes learning itself.

The book outlines the expected competencies of the effective biology teacher in terms of the following dimensions:

- Mastery of subject matter (content);
- Planning and organising lessons;
- Knowledge of the learning process;
- Fundamental teaching skills;
- Employing a variety of teaching techniques and strategies;
- Using computers in lessons;
- Working with others;
- Assessing pupils activities;
- Reviewing their own approaches;
- Selecting and using suitable resources to support the teaching and learning processes; and

- Adopting and Maintaining acceptable Personality.

Knowledge of the subject matter involves command of knowledge in the subject area to be taught and the inherent nature of such knowledge in terms of concepts, ideas, generalisations, theories, principles and laws. This entails a study of the subject matter itself and a judicious selection of the material that can be transmitted successfully to the learner. Subject knowledge is built up over time and even if the biology teacher studied the subject during the school days the teacher will need to update the knowledge. The biology teacher will also need to study new topics that have to be taught in the syllabus and this includes reading the latest subject manuals and circulars issued by the government or visiting the local teachers' resource centres and seeing what they have that may help the teacher.

Knowledge of the learning process entails a command of theoretical knowledge about learning and human behaviour as a basis for guiding the learning process. Because this kind of knowledge can be used to interpret situations and solve problems, many classroom events that might otherwise go unnoticed or misinterpreted can be recognised and resolved by applying theories of learning and concepts of human behaviour to guide the learning process. For example, the effective biology teacher will always ask the following questions about learning biology and answers provided:

- How do pupils learn biology?
- Why do pupils learn biology?
- How can we help pupils learn biology?
- How do you know when pupils are learning biology?

Attitudes to teaching are crucial components in the teaching profession. It is important for the biology teacher to display attitudes that foster learning the subject and genuine human relationships. The attitudes to teaching and learning and the relationship with class can have a marked effect on pupil attitudes and learning. Other attitudes that affect the learning process include:

- Teacher(s)' attitudes toward themselves;
- Teachers' attitudes towards peers and parents; and
- Teachers' attitudes towards the subject matter.

It pays for the teacher to be someone the pupils and other teachers admire and to always set a good example, a role model and avoids giving the wrong impression.

The fundamental teaching skills which facilitate the learning of biology include:

- The organisation of the classroom or laboratory;
- Questioning;
- Feedback mechanisms;
- Reinforcement;
- Planning and preparation for the teaching of biology;
- Set induction and closure;
- Classroom/laboratory control and management;
- Effective communication; and
- Monitoring and evaluation in the learning of biology.

Personal Practical Knowledge or Professional responsibilities include the set of understandings teachers have of the practical circumstances in which they work – beliefs, insights, habits – that enable them to do their work effectively. These include:

- Growing and developing professionally – e.g. enhancement of biology content and pedagogic skill;
- Contributing to school and community projects;
- Maintaining accurate records in the teaching and learning of biology;
- Communication with parents, colleagues, school and administration;
- Service to students, community, nation;
- Gender stereotypes – shared beliefs about the work roles or behaviour of men and women; for example 'men are better workers that women' or 'a woman's place is in the home',
- Gender bias – attitudes and practices which show gender stereotypes – e.g. showing one gender, either boys or girls, as always in the lead, active, successful and valued;
- Gender fairness – believing that boys and girls have equal value and making sure that they have equal opportunities to learn and to take responsibility; and
- Reflection – a process of self - appraisal or self-evaluation that allows the teacher 'to think above the teaching skills and practice –

e.g. evaluating strengths and weaknesses of personal teaching (SWOT analysis) of biology. Reflection should be an on-going activity, each time making notes on each lesson in the record book. It can be done alone, with colleagues or as part of an appraisal process. Reflection can help improve the teacher's quality of performance in biology and should be done regularly– e.g. what the biology teacher needs in order to improve the content of lessons or the approach to topics; the changes the biology teacher needs to make before giving other lessons on a similar topic; the methods the biology teacher needs to use for better results; improving time management; the kind of support the teacher might need to help sustain improvement.

The book intends to help the teacher identify the central issues of pedagogy that potentially impact on the teaching of biology and the related philosophies and practices employed by qualified teachers. It has been divided into 13 interrelated chapters to help the biology teacher to gain a thorough understanding of biology and how to effectively teach it. It is our contention that the information included in these chapters will promote the social transformation of the learners of biology in secondary schools and effectively prepare them for the life ahead.

The book is prepared for all those who are interested in improving their knowledge of biology education as well as teaching biology students how to learn. The book is particularly intended to help prepare biology teacher trainees in universities and colleges and to equip the practising teachers of biology with new insights into the current developments in biology education.

The book provides challenging activities intended to encourage readers to think and reflect on their practice. The readers are encouraged to attempt to work out the activities to help them extend their knowledge and skills in the teaching of biology.

ABOUT THE AUTHORS

William W. Toili is a professor of Science and Environmental Education at the Masinde Muliro University of Science and Technology, Kenya. He is a specialist in the study of Biology and Teacher Education. He is also a peer reviewer for the Commission for University Education (CUE) and a member of the Inter-University Council for East Africa (IUCEA). He has also published numerous articles in refereed journals (wtoili@yahoo.com).

Emmanuel W. Toili is an Adjunct Lecturer with St Paul's University, Kenya. He is also a PhD Researcher with the University of Nairobi, with interests in Agricultural Information and Communication Management. His focus is on enhancing the creation of an information management system that would promote efforts in harnessing biological systems to achieve food security, sustainable development and peacekeeping for all humanity. He is also working on a new book tentatively entitled *Introduction to Political Science* (emmanueltoili@gmail.com).

Mary E. M. Toili is a PhD Researcher in the Bioengineering Sciences Department at the Vrije Universiteit Brussels (VUB), Belgium. Her research interests revolve around simplifying the teaching and learning biology in schools and universities as well as the use of molecular biology tools in the dynamic agricultural world to promote sustainable development, with a strong bias in achieving food security in sub-Saharan Africa (essytoili@gmail.com).

LABORATORY: The management of the laboratory is the responsibility of the biology teacher or the head of the department

Photo: GEORGE REDGRAVE/FLICKR

1

CHAPTER ONE

SCOPE OF BIOLOGY

◆ Definition of Biology ◆ Characteristics of Living Things ◆ Key Themes and Principles in Biology

Introduction

Biology, physics, chemistry, and geology are almost always considered as sciences. This is because these subjects have evolved separate bodies of knowledge based on generalisations which can be tested or verified by repeated experimentation. This knowledge has been created through the application of the **scientific method** which is a unique process to all sciences. This chapter attempts to indicate the key and unique features of biology which form the body of knowledge of the subject. The chapter begins with the definition of the term 'biology' and then looks at the characteristics of living things that form the backbone of the study of biology. The chapter further identifies the scope of biology involving a description of the fundamental themes and principles that biologists study.

Definition of Biology

French biologist Jean Babtiste de Lamarck (1744-1829)

The term 'Biology' was coined in 1802 by the German scientist by the name Gottfried Treviranus (1776-1873). The word was derived from two Greek words-*bios* (meaning life) and *Logos* (meaning study). From the two Greek words, *Biology is the study of life*. The word was then given widespread usage by the French biologist Jean Babtiste de Lamarck (1744-1829) who saw biology as a means of encompassing the growing number of disciplines involved with the study of living things. The science of biology includes all that is known about living forms of every shape and size.

Biology is commonly defined as the study or science of life. This definition requires us to ask the question, 'what is life'? No satisfactory answer has ever been given to this commonly asked question. Dictionary definitions of life are as follows:

- Life is 'the condition which distinguishes animals and plants from inorganic objects and dead organisms'; and
- Dead is 'deprived of life'.

Then 'life' is what animals and plants have when they are not dead and 'dead' is what those same organisms are when they lack life. The dictionary definitions take us around circles and cannot get us to a clear definition of 'life'. Many biology textbooks, therefore, attempt to define biology in terms of the attributes that characterise life, namely, the characteristics that distinguish living matter from non-living matter. Biologists have formulated a list of characteristics by which we can recognise living things.

Sometimes non-living things may have one or more of life's character-

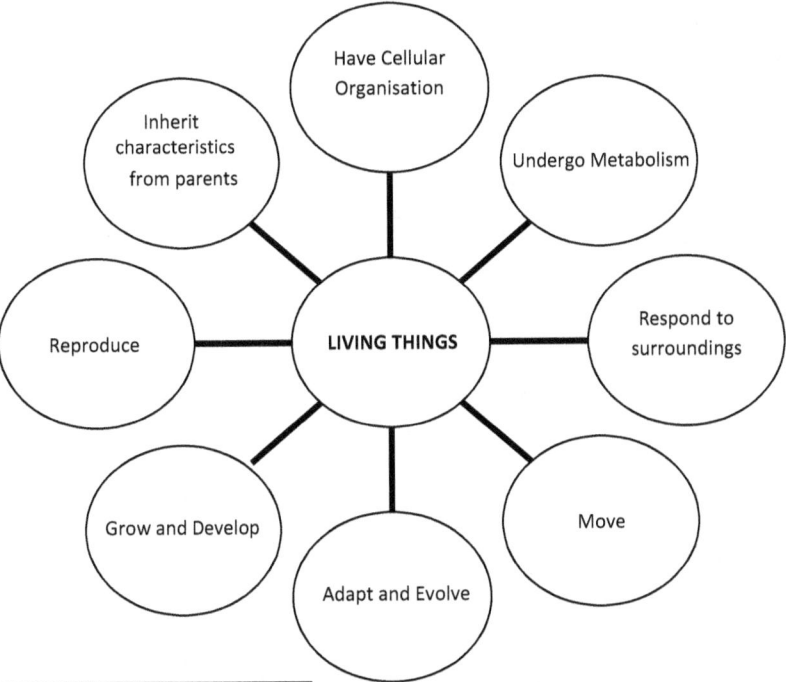

Figure 1.1: Diagrammatic Representation of Characteristics of Living Things

istics, but only when an object has all of them can it be considered as a living organism. The characteristics are described below.

Characteristics of Living Things

All living things have an orderly structure, produce offspring, grow and develop, adjust to changes in the environment and have the ability to manipulate energy and matter in a unique to manner to sustain their lives. Understanding these characteristics will help us understand how living things differ from non-living things. The following characteristics are unique to all living things (see Fig 1.1).

Living Things have Cellular Organisation

All living things show an orderly structure or organisation. All of them, whatever their form, are composed of one or more cells. A cell is a tiny three dimensional compartment with a thin covering called a membrane. Each cell contains the genetic material or Deoxyribonucleic Acid (DNA) that provides all the information needed to control the organism's life. Many organisms have a single cell (unicellular) while others have many cells (multi-cellular). In unicellular organisms such as bacteria and yeast, the life functions are carried out within a single cell.

Cell ⇒ Tissues ⇒ Organs ⇒ Systems ⇒ Organism

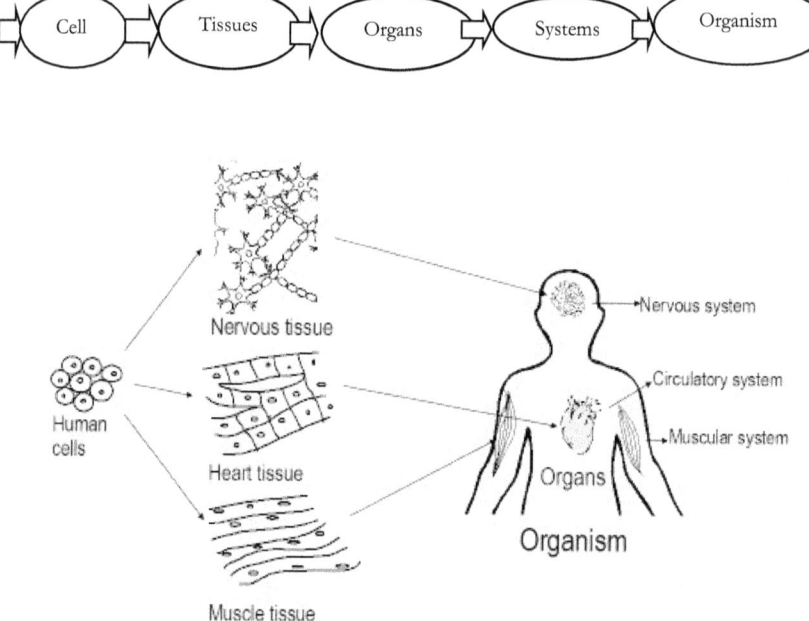

Fig 1.2 Interconnection of organs

In multi-cellular organisms such as a human being that consist of a trillion cells, the cells have specialised abilities that interact to provide the independently functioning unit called an organism. In this case, the cells are organised hierarchically to form the structures and perform the function of the organisms. Cells with similar structure cooperate with one another in units called tissues (e.g. muscles, nerves) which together serve a common purpose. Groups of tissues are organised into larger units called organs (e.g. heart) and in turn into organ systems (e.g. circulatory system). An organism consists of several organ systems that are interconnected and function together as a unit (cells, tissues, organs, systems, and organism).

However, whether an organism is unicellular or multi-cellular, all of its parts function together in an orderly, living system.

Living Things undergo Metabolism

All living things are capable of a series of chemical processes that enable them to carry out the day-to-day functions of life. The set of chemical processes is referred to as metabolism. Some of the chemical processes involve the breakdown of compounds (catabolism) while others involve the building of compounds from simpler molecules (anabolism).

A key metabolic process is a respiration by which energy is released during the breakdown of certain energy-rich compounds in the food. In this process oxygen 'burns' the energy-rich compounds to release the energy which is then transported from one place to another within cells using special energy-carrying molecules called Adenosine Triphosphate (ATP). During this process, carbon dioxide (CO_2), water (H_2O), and unusable heat are released as wastes.

The energy released is used in all life processes such as growth, thinking, movement, and circulation of fluids which require the energy. Much of the food obtained through nutrient uptake such as autotrophic and heterotrophic nutrition is used as a source of energy. The energy is stored in the chemical bonds of the complex molecules of the food substances which are broken down to release it.

Chemical wastes arising from metabolism must be eliminated from the organisms for them to continue the processes of life. The process of elimination is called excretion. Similarly, animals take in an excess of protein during nutrition and since this cannot be stored, it must be broken down (deamination) and excreted. Plants and animals are equipped with the structures of excretion.

Living Things Respond to their Surroundings

All organisms constantly interact with the major components of their environment including air, water, weather, temperature, and other organisms. All of them have the ability to respond to changes in these external environments (stimuli) and thus ensure they maximise their chances of survival. The responsive process is called irritability.

Irritability is an individual's rapid response to a stimulus, such as our response to loud noise, a dazzling light, or a hot object. Irritability occurs only in the individual receiving the stimulus and is rapid because the mechanisms that allow the response to occur, such as muscles, bones, and nerves, are already in place. For example, when we touch hot objects we quickly withdraw our hands, while the sight of food makes our mouths water. We react to these stimuli accordingly. We do the same thing repeatedly.

Although plants seem not to rapidly respond to stimuli as animals, they are more sensitive than we may imagine. They, too are affected by, and respond to stimuli, in their surroundings. The plants' response to stimuli is slower because the response requires growth or some other fundamental change in the given organism. For example, some flowers close up at night; leaves curl up when the winds blow; some leaves shrink when touched; roots grow toward moisture in the ground; while leaves respond to sunlight. It is therefore obvious that both plants and animals are equipped with structures and mechanisms for responding to various stimuli.

All organisms also maintain stable internal conditions such as level of water, minerals, temperature, and sugar. This process is called **homeostasis**. For example, our bodies remain at a temperature of around 37°C despite variations in the external temperatures.

Living Things Move

All living things move, but the degree differs from organism to organism. Animals move from place to place mainly in search of food, unlike plants which make their own food from raw materials obtained in one place. Some animals such as sponges move only at a certain stage in their life history while at other times they remain fastened on one spot. However, such animals have movement going on inside their bodies when they breathe, digest food, circulate blood and food, and excrete.

A movement occurs in plants, both within their cells and indeed within whole structures, although at a much slower rate than animals. Some bacterial and unicellular plants are also capable of locomotion.

Living Things Adapt and Evolve

Any structure, behaviour or internal process that enables an organism to respond to stimuli and better survive in an environment is called **adaptation**. Adaptations are inherited from generation to generation with some differences within any population of organisms. As the environment changes, some adaptations become more suited to the new conditions than others. Those with more suitable adaptations are more likely to survive and reproduce. Consequently, individuals with these adaptations become more numerous in the population. The gradual accumulation of adaptations over time is referred to as **evolution**.

Living Things change during their Lives (or Grow and Develop)

All living things are capable of a type of generative process referred to as growth. They take in food and change it into the living matter of their own bodies thus initiating growth. During growth living things add to their structures, repair worn out parts and store nutrients for later use. This growth process results in an increase in the amount of living matter and the formation of new structures. Growth, therefore, results in an increase in size and weight (mass) of an individual organism. For example, in the process of growth in animals, substances such as proteins contained in food are broken down during digestion then built up again (protein synthesis) before becoming part of the living animals.

In plants, simple substances drawn from the environment are built into complex chemicals which are then used to build up new protoplasm of the living plant, which increases in size (growth). Growth is not there-fore just a matter of material from the surroundings being added on to the organism as when a crystal grows. Instead, the material is changed before it becomes part of the living organisms. The crystal grows by the addition of the same materials to the outside of their surfaces. A growing tree adds each year, not a layer of carbon dioxide and water but a layer of wood and bark made from those new materials. Any young and growing animal is evidence that food is changed into the material of its living body.

Living Things Make More Living Things (Reproduce)

All living things are capable of another type of generative process referred to as reproduction that results in an increase in the number of individuals in a population of organisms. However, all living things eventually die. Although the lifespan of living organisms is therefore limited, they perpetuate life through reproduction, thereby ensuring the survival of species. The resulting offspring have the same general characteristics as the parents, whether such individuals are produced sexually or asexually. Asexual reproduction occurs when organism make identical copies of themselves, while sexual reproduction occurs when two organisms contribute to the creation of a unique, new organism. Therefore reproduction is perhaps the most remarkable property of living organisms.

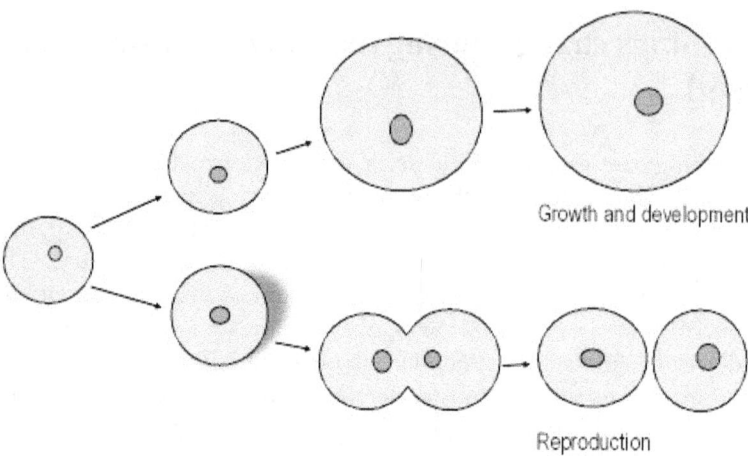

Figure 1.3: The process of reproduction and growth

Living Things Inherit Characteristics of Life from Parents

In all living things, inheritance occurs through reproduction and is controlled by the coded information contained in a long hereditary molecule called DNA. The DNA molecule is passed between organisms of succeeding generations, ensuring that the offspring are of the same species. The information that determines what an individual organism will be like is contained in a code that is dictated by the order of the subunits making up the DNA molecule, "Just as the order of letters on this page determines the sense of what you are reading" (Johnson, 2000). Each set of instructions within the DNA is called a gene. Together, the genes determine what the organism will be like. Because DNA is copied from one generation to the next, any change in a gene is also preserved and passed on to future generations. The transmission of characteristics from parents to offspring is a process called heredity and is one of the most fundamental properties of living things. Like in the case of growth, while crystals also 'grow', their growth does not involve hereditary molecules.

The Scope of Biology: Key Themes and Principles of Biology

Just as every house is organised into thematic areas such as bedroom, kitchen, and bathroom, so the living world is organised by major themes (Johnson, 2000). The themes or accepted explanations regarding living things arise out of the many tested predictions arrived at by biologists.

There are six general themes, namely:

- The diversity of life;
- Structure determines function;
- Patterns of change or evolution;
- Unity;
- Patterns of the perpetuation of life and development; and
- Interacting with the environment (Flow of energy).

Biologists Study Diversity of Life

From centuries of biological observations and inquiries, one organising principle has emerged referred to as biological diversity or ***biodiversity***. The diversity, or variety of living things, is enormous. We can appreciate the diversity of living things if we understand what a species is. A ***species*** is a group of related organisms that share common characteristics and are able to interbreed, reproducing offspring that can also reproduce. Although they are so different from one another each displays characteristics that are common among all species. The total number of plant and animal species is estimated at between 2 to 5 million; extinct species range from 15 to 16 million.

Despite the basic biological, chemical and physical similarities found in all living things, a diversity of life exists not only among and between species but also within every natural population. Biological diversity reflects history, a record of success, failure, and change extending back to a period soon after the formation of the Earth. The fact that organisms changed during prehistoric times and that new variations are constantly evolving can be verified by paleontological records as well as by breeding experiments in the laboratories.

The explanation for this diversity lies in the **theory of evolution** by natural selection. Long after Darwin had assumed that variations existed; biologists discovered that they are caused by a change in the genetic material (DNA). This change can be a slight alteration in the sequence of the constituents of DNA (nucleotides), structural alteration of a chromosome, or a complete change in the number of chromosomes. A change in the genetic material in the reproductive cells manifests itself as some kind of structure or chemical change in the offspring.

The consequence of such a mutation depends upon the interaction of the mutant offspring with its environment. The theory of evolution, therefore, forms the backbone of the study of biology just as the theory of the covalent bond is the backbone of chemistry or the theory of quantum mechanics is that of physics.

Biologists Study How Structure Determines Function

All living things have a precise structural organisation. The structures are very well suited to their functions. All living things exhibit body structures that fit their functions at every level of organisation. For example, within the cells, the shape of the proteins called enzymes that cells use to carry out chemical reactions are precisely suited to match the chemicals the enzymes must break down (Key and Lock theory). Through evolution, changes have taken place that better suit organisms to meet the challenges of living.

Biologists Study How Surface Area to Volume Ratio influences Functions

All living things exhibit body structures that not only fit their functions at every level of organisation, but such structures function best according to the surface area to volume ratio. The ratio influences the efficiency of both physiological processes and some physical mechanisms. For example, the abundance of blood capillaries within the tissues, the numerous alveoli in the lungs, the large grinding surfaces of the molar tooth, and the numerous root hairs at the tip of roots are among the characteristics intended to increase the ratio.

Biologists Study Patterns of Change

Evolution is the change in species over time. Charles Darwin in 1859 proposed the idea that this change is a result of a process called natural selection; that those organisms better able to successfully respond to the challenges of their environment become more common. This idea is summarised in one phrase: "Survival for the fittest". He suggested that 'survival for the fittest' was the basis for organic evolution or the modification of living things with time.

Darwin showed how evolution could explain the common threads underlying biological diversity. His theory contends that members of a species such as horses living today are fundamentally similar because they evolved from a common ancestor. In the case of the horse, the fossil record reveals the origin of the modern species from a succession of ancestors. Therefore, species arise through a process of 'descent with modification.'

Darwin was familiar with variation in domesticated animals and knew that varieties of pigeons could be selected by breeders to exhibit exaggerated characteristics, a process called artificial selection. The characteristics selected are passed on through generations because DNA is transmitted from parent to offspring. He visualised how selection in nature could be similar to that which had produced the different varieties of pigeons. Thus, the many forms of life we see about us on earth today, and the way we are constructed and function, reflect a long history of natural selection.

Evolution itself is a biological phenomenon common to all living things, even though it has led to their differences. Evidence to support the theory of evolution has come primarily from the fossil record, from comparative studies of structure and function, and from studies of embryological development. Evolution is, therefore, the central theme of biology because understanding how adaptations evolve by natural selection is the key to the study of life.

Biologists Study Interactions of the Environment

The study of the relationships of living things to each other and to their environment referred to as ecology, helps us understand the interdependence of living things. All organisms require energy to carry out activities. Although the sun is the original source of energy, only green plants have the ability to directly utilise this energy. Plants then serve as a source of food from which energy is derived for animals that eat them. The flow of energy in ecosystems is thus a key factor in shaping ecosystems and in affecting how many and what kinds of animals live in a community.

Biologists Study Patterns of Perpetuation of Life and Development

Although organisms are diverse in their types, they are unified in their patterns for the purpose of self-perpetuation and development. The high degree of specialisation we see among complex organisms is only possible because those organisms act to maintain a relatively stable internal environment through the process called homeostasis. Without this constancy, many of the complex interactions that need to take place within organisms would be impossible. The concept of homeostasis is today extended to include any biological system from the cell to the entire biosphere.

Whether an organism is a man or bacterium, its ability to reproduce is one of the most important characteristics of life. Because it comes only from pre-existing life, it is only through reproduction that succes-sive generations can carry on the properties of a species. The information that controls the nature and form of organisms is found in DNA. It is this chemical that is passed on from parents to offspring thus preserving the identity of species while at the same time ensuring the unique characteristics of the individual organism.

Biologists Study Unity of Life

Although organisms are diverse in their types they are unified in their patterns. Whether an organism is unicellular or multi-cellular, all of its parts function together in an orderly, living system. They have certain biological, chemical and physical characteristics in common. For example, all organisms:

- Are composed of some basic units or cells;
- Are composed of the same chemical substances. For example, all have DNA in the form of genes which account for the ability of all living matter to replicate itself exactly and to transmit genetic information from parent to offspring; and
- Have similar basic functioning. This is because the action of an organism is determined by the manner in which its cells interact and since all cells interact in much the same way their functioning is similar.

All organisms have similar basic functioning. This is because the action of an organism is determined by the manner in which its cells interact and since all cells interact in much the same way their functioning is similar. They have a precise structural organisation whereby the structures are suited to their functions. For example, within the cells, the shape of the proteins called enzymes that cells use to carry out chemical reactions are precisely suited to match the chemicals the enzymes must break down or manipulate. Similarly, within the organisms, body structures are carefully designed to carry out their functions. Life has existed on Earth for over three billion years, a long time for evolution to favour changes that better suit organisms to meet the challenges of living.

Biology Studies about Application of Technology to Biological Systems (Biotechnology)

Biotechnology is the application of technology to biological systems, living organisms, or derivatives thereof, to make or modify products or processes for specific use. Simply put, biotechnology is the use of biological processes, organisms, or systems to manufacture products intended to improve the quality of human life. This involves manipulation as through genetic engineering of living organisms or their components to produce useful usually commercial products such pest-resistant crops, new bacterial strains, or novel Areas of interest where this technology is applied are in medicine, agriculture, and criminology.

In medicine, biotechnology is used to get organisms to produce new drugs or using stem cells to regenerate damaged human tissues and perhaps re-grow entire organs. In the manufacture of drugs for various diseases including diabetes, HIV control, and high blood pressure control, among others. Insulin is produced through modern biotechnology to help diabetes patients to cope with the problem. DNA matching is now being used to solve paternity problem. In industry, it is used to get industrial processes such as the production of new chemicals or the development of new fuels for vehicles

In agriculture, several types of pest-resistance genes have been transferred to plants to make them resistant to pests. In addition, the genetically modified organisms are being used to improve agricultural yield. Development of pest-resistant grains accelerated the evolution of disease-resistant animals. Or even controlling proliferation of noxious waterborne organisms, and improved species of plants and animals.

In criminology, crime detection employs medical principles to detect crime, for example, the use of fingerprints to detect suspects in a crime. The pattern of DNA determined in this instance is prepared from specimens collected at the scene of a crime.

Conclusion

In conclusion, it can be said that the eight characteristics of living things are only the observable characteristics that guide us to determine if an object before us is living matter or non-living matter. In addition, all living things possess cells in which occur essential activities such as metabolism, maintenance of stable internal conditions, and creation of hereditary molecules.

Every living organism has a set of instructions contained in its genes and uses DNA to transmit hereditary information to offspring and to direct its metabolism, organisation, and reproduction. The investigation of all these attributes helps us understand life as a complex and dynamic process that is constantly changing. In a nutshell, biology is a unified group of life sciences dealing with development, growth, metabolism, response, reproduction, evolution and interrelations of all living things. It deals with all of the physiochemical aspects of life.

References

Enger, Eldon D, and Ross, Fredrick C (2000) *Concepts in Biology*, 9th Edition. Boston: McGraw Hill

Johnson, George B. (2000). *The Living World*, 2nd Edition Boston: McGraw Hill.

Raven, Peter H., and Johnson, George B. (1999) *Biology*, 5th Edition. Boston: McGraw-Hill.

DNA IN MOTION: To understand biology as a science, we must therefore first understand the nature of the scientific process

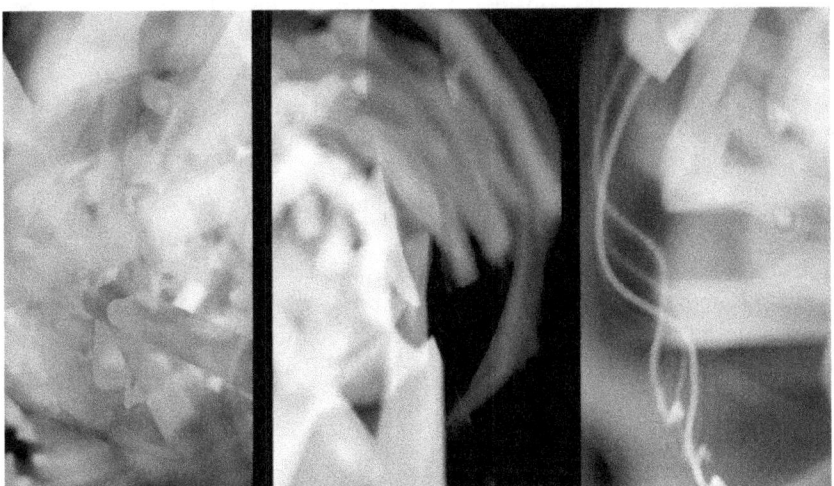

Photo: MARC SAMSOM/FLICKR

2

CHAPTER TWO

BIOLOGY AS A SCIENCE

◆ The Scientific Process ◆ Fundamental Characteristics of Science ◆
Why Biology is considered as a Science

Introduction

Biology is almost always considered as a science. This is because the subject has evolved a distinct body of knowledge based on generalisations which can be tested or verified by repeated experimentation. This knowledge has been created through the application of the scientific method, which is a unique process for all sciences. Biology originated from a twofold urge: to understand the living matter and to make use of nature. Understanding nature relies on the discovery of fundamental truths which is accompanied through basic research that chiefly utilises the scientific method. Application of knowledge about nature for human use in medicine, agriculture, and industry is achieved through applied research.

To understand biology as a science, we must therefore first understand the nature of the scientific process. This chapter attempts to indicate the nature of biology within the broad framework of the discipline of science.

The Scientific Process: The Unifying Theme of all Sciences

The term 'science' is derived from the Latin word *'scientia'*, meaning 'to know'. Science may thus be viewed as a discipline of knowledge. According to Webster's Dictionary, science is defined as a systematised knowledge derived from observation and experiment carried out to determine the principles underlying what is being studied. Science is a particular way of inves-tigating the world, of forming general rules about why things happen by observing particular situations. In other words, science is a particular way of investigating the world in order to build an understanding of the natural world. A scientist is an observer, someone who looks at the world in order to understand how it works.

All scientists study the natural world and go about their work in a similar way by using the **scientific method**; a product of thought. The scientific method is a process or a set of procedures scientists use to answer the questions they ask about the natural world.

All scientists ask questions that can be investigated by means of experimentation. They ask questions that can be tested, producing results that are open to verification by others in their discipline. They don't ask philosophical or religious questions such as "What is the meaning of life," "Does life exist after death?" Such questions cannot be tested. Biologists pose questions that help them identify and understand those things that influence the process of life. They are concerned with the exploration of the behaviour and properties of animate material or living matter. Physicists are concerned with the exploration of energy and general properties of inanimate materials and how they interact with other forces in the environment, while chemists are concerned with the particulate properties of inanimate materials.

We seek to learn more about the living matter through the science of biology. We do this by seeking evidence by observation and subjecting the evidence to the criteria of the scientific method. The ensuing evidence is then reduced to fundamental generalisations that make up biological knowledge.

Stages in Scientific Method

The scientific method can be said to have six stages:

- Observation – observing what is going on;
- Hypothesising – forming a hypothesis;
- Experimentation – testing hypotheses or predictions and controlling variables;
- Interpretation - explaining what the data say;
- Conclusion – eliminating one or more or the hypotheses and making generalisation; and
- Dissemination of results.

The stages of the scientific method are as summarised in Figure 2.1. The stages are described next.

Observation

The scientific method usually starts with observations that lead to questions. Observation is one of the most important processes of science. For example, a Form Three biology student notices a variation in the growth of climbing ornamental plants (Zebrina) in the project conducted by his classmates. He observes that zebrina plants grew healthy when given regular doses of Di-Ammonium Phosphate (DAP) fertilizer. He has also read the literature stating that DAP fertilizer caused ornamental plants such as Zebrina to grow taller and healthier. These observations would trigger the question: *what caused ornamental plants (Zebrina) to grow healthier?*

Hypothesis Formulation

After making careful observations concerning a particular area of science such as biology scientists construct a hypothesis, which is a suggested explanation that accounts for those observations. A hypothesis is an educated guess that tentatively answers the question raised. It is a prediction of what may happen in the investigation. Before scientists formulate a hypothesis, however, they gather information regarding their questions. Only after searching the scientific literature and synthesising all the information they can find regarding their questions do they generate a hypothesis. The hypothesis, therefore, is not just a guess but a well thought out prediction based on available knowledge and generalisations made from observations.

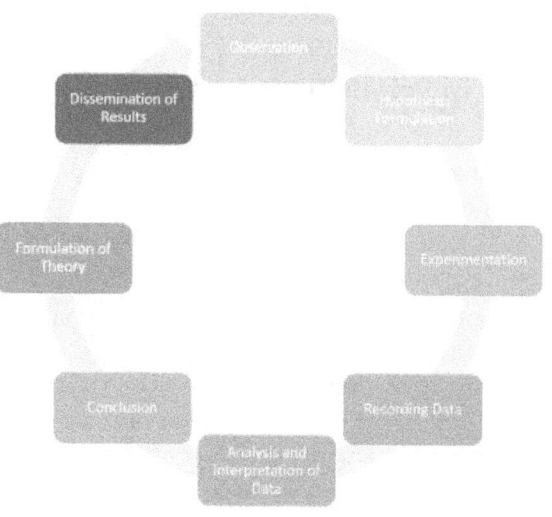

Figure 2.1: A Diagrammatic representation of the Scientific Method

This pattern of thought in which generalisations are developed from specific instances is called **Inductive Reasoning**. Suppose in the case of the Form Three student, he was told by his friends that one key to growing healthy ornamental plants is giving them regular doses of DAP fertilizer and had also read the literature stating that DAP fertilizer caused ornamentals to grow taller and produce healthier greener plants. From these observations and information, he would make the inductive generalisation that DAP fertilizer helps ornamental plants grow well. This generalisation becomes a hypothesis when stated in an "**if……..then………**" format.

The student's hypothesis would be stated as follows: "If Zebrina plants are given DAP fertilizer, then they will grow taller and be greener than plants with no fertilizer."

More often than not, scientists will form alternative hypotheses if they have more than one guess about what they observe.

Experimentation (Testing the Hypothesis)

A hypothesis is usually a prediction that can be tested. A prediction is what you expect to happen if a hypothesis is true. For example, the student's hypothesis predicts that if DAP fertilizer is responsible for producing taller and healthier ornamental plants, then it should be possible to detect these effects in Zebrina plants which receive regular doses of DAP fertilizer.

Scientists test hypotheses by attempting to verify some of its predictions. We call the test of a hypothesis an experiment. In an experiment set to test the hypothesis, we must first analyse the hypothesis. In analysing the hypothesis, the factor we are manipulating (e.g. fertilizer) is stated after the "if" part of the hypothesis. This factor is the **independent variable** (or manipulated variable). The term variable refers to a factor that varies or changes. To test his hypothesis, the student needed to add different amounts of DAP fertilizer to Zebrina plants to determine the effects on plant growth and leaf colour. The factor stated after the "then" part of the hypothesis depends on the manipulated (independent) variable. This factor is referred to as the **dependent variable** (or responding variable). It varies in response to changes in the independent variable.

Scientists test hypotheses by means of **controlled experiments**. An experiment tests a hypothesis by producing data that either support or refute the prediction made in the hypothesis. This "testing" process involves the use of **deductive reasoning.** The deduction begins with a general statement (the prediction about DAP fertilizer and its effects on Zebrina plants) and proceeds to a specific statement (the effects of DAP fertilizer on Zebrina plants). In a controlled experiment, the independent variable is manipulated and changes in the dependent variable are observed while all other possible variables are kept constant; they are **controlled**. The variable to be kept constant include any factors in the experimental set-up that can change such as type of soil, the pH of the soil, the amount of moisture in the soil, the amount of time the plants are exposed to light, the temperature in which the plants are grown, and the type of plant used, among other factors.

As we design a controlled experiment to test the hypotheses, we should first outline the way in which we would manipulate the independent variable. This plan should include the treatment and the control. In performing a controlled experiment one usually carries out two parallel tests, one the experimental, the other the control. The control is the standard against which the treatment can be compared. Any changes in the control are the result of factors other than the treatment. Controls make it possible to draw clear-cut conclusions from the results of experiments.

Over time pre-selected observations are made in the experiment and data are collected such as leaf colour and vine length and the data are recorded appropriately. The latter is quantitative data because these data are based on numerical measurement. The data regarding leaf colour are called qualitative data because these data are descriptive and not based on numerical measurements.

Recording of Data

Over time observations made in the experiment generate the data that are collected such as leaf colour and vine length and the data are recorded appropriately. The latter is quantitative data because these data are based on numerical measurement. The data regarding leaf colour are called qualitative data because these data are descriptive and not based on numerical measurements. Suitable methods of data presentation are used such as tables, pie charts, and graphs among others.

Analysis and Interpretation of Data

A valid interpretation of results is made based on the recorded quantitative and qualitative data. The interpretation is guided by the qualitative and quantitative data regarding the independent, dependent and control variables.

Conclusion

From the data obtained it is necessary to ask: Do the data support the hypothesis: "*If I add DAP fertilizer to the soil of my Zebrina plants, they will grow taller and be greener than plants with no fertilizer.*"?

Suppose the qualitative data indicated that the plants with fertilizer were greener than the control with no fertilizer, while the quantitative data suggested that DAP fertilizer helped Zebrina grow greener and showed a greater increase in growth. The student would be right to accept the hypothesis and to make conclusions.

Scientists usually perform experiments over and over again (repeated trials) before drawing conclusions from their data. Otherwise, it would be risky to generalise from the results of only one or a few experiments. We can only draw firm conclusions only after repeating experiments obtaining consistent results. Conclusions are usually in the form of generalisations such as theories, principles, laws, and hypothesis.

Theory Building

A hypothesis that has been tested and not rejected is tentatively accepted. As scientists repeat or replicate each other's work, their data may uphold hypotheses again and again. The related hypothesis that stands the test of time (their predictions are often tested and never rejected) are sometimes crafted into general statements called **theories**. A theory is a unifying explanation for a broad range of observations. Theories are broad statements which are a result of years of observation, questioning, experimentation and analysing data. Thus we speak of the theory of gravity, the theory of evolution, and the atomic theory.

A theory is a synthesis of hypotheses that have withstood the test of time and is, therefore, a powerful concept that helps scientists make dependable predictions about the natural world. Theories are supported by such an overwhelming weight of evidence that they are accepted as scientifically valid statements. However, the future evidence may cause a theory to be revised, since scientific knowledge grows and changes as new data are collected, analysed and then synthesised with information that came before. A scientist's acceptance of a theory is always provisional and scientific knowledge is usually tentative. If a theory has been proven beyond reasonable doubt to be true, then the theory becomes **a law**. Examples of theories in biology in biology include the following:

- **The Germ Theory of Disease**: it states that certain diseases, called infectious diseases, are caused by micro-organisms that are capable of being transmitted from one individual to another;
- **Cellular Theory of Life**: states that living things are made of cells; and
- **Theory of Evolution**: it states that species adapt to their environments and those that are most fit survive.

The following are examples of laws in biology:

- **Biogenetic Law**: it states that all living things come from pre-existing living things. It is broader than a theory; and
- **Law of Segregation**: it states that during reproduction, the two factors that control each trait separate or segregate, with one factor from each pair passed to the offspring.

Dissemination of Results

Scientists finally publish the results of their work in journals or other authoritative sources for other rated scientists to debate and verify the results. The results are contained in what Medawar (1969) calls the 'scientific paper'. The paper outlines:

- **Introduction** – in which the scientist merely describes the general field in which his/her scientific talents are going to be exercised;
- **Literature review** – in which the scientist outlines the previous work done on the problem of study. The scientist concedes that others have dimly groped towards the fundamental truths that they are now about to expound;
- Results – which consists of a body of factual information in which it is considered bad to discuss the significance of the results one is getting. One has to pretend that their mind is unconcerned about this information. One reserves all appraisal of the scientific evidence until the 'discussion' section; and
- Discussion – in which the scientist pretends to ask oneself if the information collected actually means anything; or if any general truths are going to emerge from the completion of all the evidence outlined in the 'results' section. The scientist then makes conclusions by declaring the knowledge discovered.

The conception underlying this type of scientific reporting is that scientific discovery is an inductive process in which scientific discovery or the formulation of the scientific theory, starts with the utilisation of the senses. It starts with simple observation and out of this sensory evidence, embodied in the form of simple propositions or declarations of fact, generalisations will gradually emerge.

Limitations of Science and the Scientific Method

Scientific progress is not just the result of applying a series of steps called the scientific method. In other words, trial and error testing do not inevitably lead one to discover scientific knowledge. If this were indeed true, a computer would make a good scientist. The scientific method is just an ideal model of how science works and is not always possible to adhere rigidly to the model (i.e. steps of the scientific method).

A scientist does not follow a fixed method – the scientific method – to form hypotheses but relies also on judgment and intuition. A scientist has to design experiments with a pretty fair idea of how they will come out. A hypothesis that a successful scientist test is not just any hypothesis; it is a 'hunch' or educated guess in which one integrates all that they know. This is because insight and imagination play such a large role in scientific progress that some scientists are so much better at science than others.

Science, on the other hand, is limited to the study of organisms and processes that we are able to observe and measure. Supernatural, religious and unexplained phenomena are beyond the realm of scientific analysis because they cannot be scientifically studied, analysed or explained. This is because scientists in their work are limited to objective interpretations of observable phenomena.

Examples of How Biologists Work using the Scientific Method

a) A Class Biology Project

Njoroge observed that many of his Third Form friends in the high school grow bean seedlings in their plots in the school garden that looked green and healthy. Since he struggled to keep his alive, he asked them their 'secret'. Many told him that one key to growing healthy bean seedlings was giving them a regular supply of Di-Ammonium Phosphate (DAP) fertilizer. He had also learned in class that DAP fertilizer caused plants to grow taller and produce healthier and greener plants.

From these observations and information, Njoroge made the inductive generalisation that DAP fertilizer helps bean plants grow well. He then made the following hypothesis: *'If bean seedlings are given DAP fertilizer, then they will grow taller and be greener than plants with no fertilizer.'*

To test his hypotheses, Njoroge added different amounts of DAP fertilizer (using bottle tops/caps) to bean seedlings to determine the effects on plant growth and colour. He first measured the total length of the stems and observed the leaf colour and repeated this after five weeks. He recorded the data as shown in Tables 2.1 and 2.2:

Table 2.1: Total Stem Length in cm

Amount of fertilizer (capfuls)	0 cf	½ cf	1 cf	1½ cf	2 cf
Length of stem (0 weeks)	6	7	9	10	11
Length of stem (5 weeks)	13	15	18	19	22
Increase in stem length	7	8	9	9	11

Key: cf = capfuls of DAP fertilizer

Table 2.2: Description of Leaf Colour on Selected Bean Seedlings

Amount of fertilizer	0 cf	½ cf	1 cf	1½ cf	2 cf
Leaf colour (0 weeks)	Mg	Mg	Mg	Mg	Mg
Leaf colour (5 weeks)	Lg	Lg	Dg	Dg	Dg
Change in Vine colour	-ve	0	+ve	+ve	+ve

Key: mg - Medium green
    Lg - Light green
    Dg - Dark green
    +ve - Positive change (turned darker)
    -ve - Negative change (turned lighter)
    0 - No change

Njoroge studied the data to see if they supported the hypothesis: *'If I add DAP fertilizer to the soil of my bean plants, they will grow taller and be greener than plants with no fertilizer.'*

He noticed that plants with fertilizers were greener than the ones without fertilizers. One capful to two capfuls produced greenest bean plants. Njoroge was convinced that the data supported his earlier stated hypothesis. However, scientists perform experiments over and over again (i.e. **repeated trials**) before drawing conclusions from their data.

From the story, it is evident that Njoroge used the scientific method, a process that scientists use to answer questions about nature.

b) Edward Jenner and the Control of Small-Pox (Adapted from Enger and Ross, 2000)

Edward Jenner first developed the technique of vaccination in 1795. This was the result of a 26-year study of two diseases: cowpox and smallpox. Cowpox was known as **vaccinae**. From this word evolved the present term vaccination and vaccine.

Jenner observed that milkmaids developed pock like sore after milking cows infected with cowpox, but they rarely become sick with smallpox. He asked the question. *'Why don't milkmaids get smallpox?'* He developed the **hypotheses** that the mild reaction milkmaids had to cowpox protected them from the often fatal smallpox. This led him to perform an **experiment** in which he transferred pus-like material from the cowpox to human skin and discovered that **vaccinated** people were protected from smallpox.

Edward Jenner (1749-1823)

When these results became known, it sparked a mixed public reaction. Some people thought that vaccination was the work of the devil. Many European rulers supported Jenner by encouraging their subjects to be vaccinated. Napoleon and the Empress of Russia were very influential and, in the United States, Thomas Jefferson had some members of his family vaccinated. Many years later, following the development of the germ theory of disease, it was discovered cowpox and smallpox are caused by viruses that are very similar in structure. Exposure to the cowpox virus allows the body to develop immunity against the cowpox virus and the smallpox virus at the same time. In 1979, almost two hundred years after Jenner developed his vaccination, the Centres for Disease Control and Prevention (CDC) in the United States and the World Health Organization (WHO) of the United Nations declared that smallpox was extinct.

From Jenner's story, new knowledge in biology was generated as follows:

- **Making a decision** about what question to ask that elicits the right information;
- **Exploring alternative resource** to determine if the question raised is a good one or if others have already explored it.
- **Formulating the hypothesis** - posing a possible answer that is testable and accounts for all the known information. A hypothesis is a statement that provides a possible answer to a question or an explanation for an observation.
- **Experimenting** to allow testing of the hypothesis using control and experimental variables; collecting and analysing the data. An experiment is a recreation of an event or occurrence in a way that enables a scientist to support or disprove hypotheses. During experiment new information come about prompting new questions to be raised that can lead to even more experiments; and
- **Formulating a theory:** Biologists repeat experiments and share information with others over a long period of time. If information continues to be considered valid and consistent with other closely related research, biologists will recognise that a theory has been established or that the information is consistent with existing theories. Truth is then pronounced.

Fundamental Characteristics of Science

There are several characteristics of science which are different from other disciplines. These include the following:

Science Studies about the Real World

Science is knowledge only about the real world vested in related concepts and conceptual framework. Thus, science investigates the real or natural world; it studies systematically the natural objects, phenomena, and events that are observable through our senses or extensions of such senses. This is because science presumes that the things, phenomena, and events in the natural world occur in consistent patterns that are comprehensible through careful systematic study and that the basic rules everywhere in the universe are the same (consistency).

For example, scientists believe that everything in the universe is composed of the same elements as those found on earth. Science does not study those things that are not observable using the senses or the extensions of the senses. Such things are believed to be metaphysical and are best dealt with under other forms of knowledge. For example, science does not have the means to settle issues of good and evil, or beliefs such as the existence of supernatural powers and beings, or the purpose of life.

Scientific Knowledge is Empirical

Because science investigates the real world it relies on observation and experimentation as the basis of evidence upon which truth is pronounced as scientific knowledge. Thus scientific knowledge is empirical and rejects supernatural explanations for observed phenomena. Opinion is not as acceptable as scientific knowledge. The humanities disciplines do not rely on experimentation for the formulation of statements. Supernatural explanations and opinion may count as knowledge in such disciplines.

Scientific Knowledge is Tentative

Science is a process for producing knowledge. The process depends on making careful observations of phenomena and on inventing theories for making sense out of those observations. Scientific knowledge does not, therefore, claim to be absolute truth and is subject to change in the wake of new evidence. This is because most scientific work tries to disapprove hypotheses rather than prove them correct. Change in scientific knowledge is therefore inevitable because new observations may challenge prevailing theories. If over a period of time, no one succeeds in disapproving a given hypothesis, it becomes increasingly accepted as a theory or later as law.

Science uses the Scientific Method

Science is a discipline that employs the scientific process or the scientific method to find out about the real world in a systematic manner. Through this process, people have gradually built up a framework of scientific knowledge-a set of ideas and phenomena that enable us to interpret the world in a characteristic way. The whole scientific process is a search for empirical evidence based on observation, hypothesis formulation, and experimentation. Only truth established through testing is accepted to build up scientific knowledge.

Scientific Knowledge is a Product of Logical Reasoning

The scientific method relies on logical reasoning in creating scientific knowledge. Logical reasoning is of two types: Deductive Reasoning and Inductive Reasoning.

Deductive Reasoning

Deductive reasoning (Fig. 2.2) is the type of reasoning where general principles are applied to make specific decisions, observations, or to predict specific results. A deduction is actually a conclusion drawn from general principles. All in all, deductive reasoning is used to test the validity of general ideas in all branches of knowledge. General principles are constructed and then used as the basis for examining specific cases to which they apply.

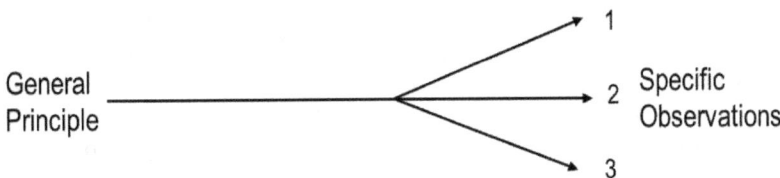

Figure 2.2: A Diagrammatic representation of Deductive Reasoning

Deductive Reasoning is therefore used both in scientific and non-scientific investigations. In terms of its use in non-scientific investigations, it is the reasoning applied in mathematics, philosophy, political science, and ethics; it is also the way a computer works. In our day to day decisions, we also apply deductive reasoning. For example, when you want to know how to get to Malindi from Mombasa, you do not conduct a scientific investigation – instead, you look at a map to determine a route. Making individual decisions by applying a 'map' of accepted general principles is called **Deductive Reasoning**.

In scientific investigations, deductive reasoning is usually applied during the testing process (experimentation). The deduction begins with a general statement (e.g. the prediction about DAP fertilizer and its effects on houseplants as set up in the hypotheses) and proceeds to a specific statement (the effects of DAP fertilizer on houseplants as observed from experimental data).

Inductive Reasoning

Inductive Reasoning (Fig.2.3) is the reasoning where specific observations are used to construct general scientific principles. These are derived from

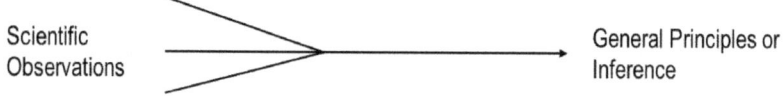

Figure 2.3: A diagrammatic representation of inductive reasoning

observations of the physical world around us. Scientists discover the general principles that govern the operation of the physical world by carefully looking at the world to understand how it works. It is from their observations that scientist determines the general principles that govern our natural world. Indeed science is best defined as the determination of general principles from observation and experimentation. This underlies the role of inductive reasoning in science.

Inductive Reasoning is key in the formulation of hypotheses during scientific investigations. In formulating a hypothesis, scientists first gath-er information regarding the questions raised. Only after searching the scientific literature and synthesizing all the information they can find regarding their questions do scientists generate hypotheses. The hypothesis, therefore, is not just a guess but a well-thought-out prediction based on available knowledge and generalisations made from observations. This pattern of thought by which we develop generalisations from specific instances or observations is called **Inductive Reasoning**.

Inductive Reasoning is key in the formulation of hypotheses during scientific investigations. In formulating a hypothesis, a scientist first gathers information regarding the questions raised. Only after searching the scientific literature and synthesising all the information they can find regarding their questions do scientists generate hypotheses. The hypothesis, therefore, is not just a guess but a well-thought-out prediction based on available knowledge and generalisations made from observations. This pattern of thought by which we develop generalisations from specific instances or observations is called **Inductive Reasoning**.

Inductive reasoning first became popular and important to science about 400 years ago in Western Europe when scientists like Isaac Newton and Francis Bacon began to conduct experiments and from the results, they implied the general principles about how the natural world operated. For example, Isaac Newton conducted numerous simple investigations from which specific observations of the falling apple led to a general principle: *all objects fall toward the centre of the earth*. This became the law of gravity which we use today. Newton thus constructed a mental model of how the world worked. Like Newton, scientists today formulate hypotheses and build general models based on observations.

Scientific Ideas are Quantifiable

Scientific ideas can be quantified. In this way, scientists are able to answer questions such as: 'How long? How far? How much?' To get answers to these questions, measurements are conducted and numbers are used. This helps in describing experiments conducted and in formulating conclusions and generalisations that can be tested with precision and understood by others throughout the world. This is one reason why mathematics is an integral part of science.

Mathematics is sometimes described as the language of science. Without mathematics, there would be no science; there would be no Newton's law of gravity, no electronic computers, no atomic energy, and no conquest of space. This is mainly because of mathematics:

- Enables scientists to describe experiments conducted and conclusions reached in a way that can be understood by other scientists throughout the world;
- Enables scientists to observe accurately the objects, forces, and changes that occur in the universe using measurements and numbers; and
- Allows scientific observations to be quantified, thus enabling scientists to answer questions involving numerical values, such as: How much? How long? How far? How heavy?

Science is Parsimonious

Science prefers a simpler explanation of phenomenon and events, using as few words as possible, as opposed to the more complex explanation. For example, Darwin's theory of evolution is reduced to the simple and very descriptive phrase, 'survival for the fittest'. Einstein's general theory of relativity has been reduced to the simple equation, $e=mc^2$.

Science is Honest

Science is honest as it relies on a characteristic way of testing findings intended to become science through experimentation. The results are published in scientific journals so that other scientists verify them. In this way, there has been continuity and stability in scientific knowledge. In other words, honesty leads to durability and improvements in scientific knowledge.

Why Biology is considered a Science

Biology is therefore considered as a science because of the following characteristics:

- It is a unified body of knowledge about living things (real world) and how they interact with their surroundings that has been accumulated over several centuries and which consist of fundamental generalisations that are derived in a logical way;
- Biology creates its knowledge by subjecting the empirical evidence to the rigours of the scientific method in the process of formulating the new biological knowledge;
- In deriving the biological knowledge inform of generalizations, biologists employ logical reasoning in terms of deductive and inductive reasoning at the various stages of the scientific process;
- The generalisations (biological knowledge) that have over the years been accumulated are verifiable as they can be tested by repeated experimentation. This is because empirical evidence is capable of being verified or disapproved by further observation. If the event occurs only once or cannot be repeated in an artificial situation it is impossible to use the scientific method to gain further information about the event and explain it;
- Biology relies on empirical evidence in formulating new biological knowledge. In this process, biologists use direct observation inform of their senses or an extension of their senses (e.g. microscopes, tape recorders, x-ray machines, thermometers, indicators, etc) to record information that forms empirical evidence;

- We seek to learn more about nature through the science of biology. We do this by seeking evidence by observation and subjecting the evidence to the criteria of the scientific method. The ensuing evidence is then reduced to fundamental generalisations (biological knowledge); and
- Biology, like other sciences, originated from a twofold urge: to understand nature and to make use of nature. Understanding nature relies on the discovery of fundamental truths which is accomplished through basic research. Application of knowledge about nature for human use in medicine, agriculture, and industry is achieved through applied research.

Box 2.1

Discussion Questions

1. Early in the 20th century, some people believed that Biology was not a natural science and could not be compared to the hard sciences like Physics and Chemistry. What advice would you give to those who might still believe such people?
2. Why should children in schools study Biology?
3. Explain why Biology does not always rely on inductive reasoning in the formulation of new knowledge.

Conclusion

Biology is one of the areas of knowledge within the field of science. It relies on empirical evidence in formulating new knowledge. In this process, biologists employ the scientific method that incorporates direct observation inform of their senses or an extension of their senses (e.g. microscopes, tape recorders, x-ray machines, thermometers, indicators, etc.) to record information that forms empirical evidence. In addition, biologists formulate hypotheses as a basis for conducting experiments and collection of data from which theories are constructed and disseminated. Biology also shares several other characteristics of science such as honesty, parsimony, quantification, among others. These are some of the key features that biology learners in schools should be encouraged to develop so that they are able to do science as scientists do.

References

Enger, Eldon D, and Ross, Fredrick C (2000) *Concepts in Biology*, 9th Edition. Boston: McGraw Hill

Johnson, George B. (2000) *The Living World*, 2nd Edition Boston: McGraw Hill.

Raven, Peter H., and Johnson, George B. (1999) *Biology*, 5th Edition. Boston: McGraw-Hill.

TEACHING SECONDARY SCHOOL BIOLOGY

BIOLOGY EDUCATION attempts to answer the fundamental professional questions concerning the teaching and learning of biology

Photo: PETR SMELC/FLICKR

3

CHAPTER THREE

NATURE AND GOALS OF BIOLOGY EDUCATION IN SECONDARY SCHOOLS

◆ Nature and Meaning of Biology Education ◆ Why Teach Biology ◆ Outcomes of Biology Education

Introduction

The study of biology in secondary schools in most developing countries, Kenya included, aims at providing the learners with the requisite biological knowledge as well as scientific skills and attitudes which should enable them to transform the environment for their own benefit and that of families and communities. The focus is on understanding biological processes which require the application of higher thinking skills such as theorising, applying knowledge to new situations, explaining, raising questions, hypothesising, and evaluating alternative ideas or explanations. The curriculum is therefore broad-based in nature and anchored on biological concepts and principles as well as scientific skills and attitudes.

This chapter sheds light on the nature of biology education in secondary schools and further outlines the goals and objectives of the biology curriculum.

Nature and Meaning of Biology Education

Biology education is a component of the wider discipline of science education. Biology education refers to that part of science education that deals with active transmission of knowledge about the living matter as well as the associated scientific skills and attitudes to the learners not traditionally considered part of the scientific community and involves pedagogical underpinnings that the teacher must incorporate. Biology education attempts to answer the fundamental professional questions concerning the teaching and learning of biology outlined in Table 3.1. The solutions to these questions are mostly found from research findings. The greatest input comes from research in theories of learning provided in professional areas including educational psychology, philosophy of education, educational technology, sociology of education, curriculum development and educational evaluation among others. If one is trying to provide answers or solutions to these questions then one is studying **biology education**.

Table 3.1: Questions that biology education must always answer

Key Question	Area of education providing the answer
Why should biology be taught to the learners?	Philosophy of science
What content and skills of biology should be taught at a particular age of the learners?	Psychology of education / Curriculum development
What is the most appropriate sequence of presenting biological knowledge in a course or lesson?	Curriculum development / Psychology of education
Who should learn biology and why?	Philosophy of education
How is biology taught and learned?	Educational technology

Who should teach biology and why?	Sociology of education
What qualities constitute a good biology teacher?	Sociology of education
What methods of teaching biology produce the desired results?	Psychology of education
What type of learning resources will produce more permanent learning in biology?	Psychology of education / Educational technology
How would we know that learners have learned biology?	Educational evaluation
How is the financing of biology teaching and learning going to be done?	Sociology of education
What is the future of biology in society and with other disciplines?	Philosophy of science

Why Teach Biology?

To study an enormous subject like biology without good reasons would be a real waste of time. However, there are many good reasons for studying the science of biology in schools.

Biology helps us to Improve Quality of Health

Many years ago, a lot of people died from diseases such as smallpox, diphtheria, and measles. Through research, biological knowledge has been produced and applied to control such diseases. Today, we rarely encounter these diseases. Scientists are also constantly finding causes of, and cures for, the diseases such as cancer, polio and HIV and AIDS. These diseases, too, will eventually be conquered. Biologists have also helped to increase our lifespan. Through the study of biology, we learn more about the structure of the human body and the working of its various parts. This understanding helps us to learn how to take better care of ourselves than before.

Biology helps us Produce more Food

Unlike our ancestors who relied on hunting, gathering, and fishing for food, today we must cultivate plants and rear animals to supply the food we need. To do this successfully, we must have a better understanding of plants and animals and how best we can depend on them for food and controlling the diseases that affect these organisms.

Biology helps us Conserve our Natural Resources

Our forests and other natural resources have been carelessly destroyed in many parts of the world. As our populations have continued to soar, this has led to many environmental problems such as desertification, reduced biodiversity, and extinction of rare species of plants and animals. Biology helps us to understand the interdependence of plants, animals and the inanimate environment as a basis for the conservation of the environment in total.

Biology helps us to be Better Citizens

Man is a social animal and is prone to diseases spread from others and the environment. This creates health problems. To prevent the spread of diseases, public health measures are preferred. Health regulations focusing on proper sewage disposal, water supply, and preventive medicine are required. These are social as well as biological problems. If every citizen is educated in the fundamentals of biology then they can participate effectively in the management of their health and safety matters.

Biology helps to Prepare for Careers

A vocation is a chosen life work. The study of biology is a requirement for many vocations such as medicine, agriculture, ecologist, florist, nursing, landscape architect, farmer, teacher, bacteriologist, dentist, biotechnologist among others. Since we are living organisms, every person is using biology from birth to death. How long one lives may to a large extent be determined by how effectively biological knowledge is applied. People who have studied biology have a great advantage over those who have not. Biology is thus one of the most practical subjects you will ever study.

Biology helps us to enjoy our Leisure Time

Even if you enter vocations that don't need biology as a requirement, knowledge of biology opens up new avenues for recreation. On such occasions, you will find the world around you interesting, such as trees, flowers, birds, insects, and wild animals such as leopards, lions, snakes. You may want to take pictures of such beautiful objects or sceneries such as rivers and valleys. There are also many hobbies you may engage in that depend on biological knowledge such as fishing, raising pets, gardening, photographing living things, making and keeping aquaria and herbaria.

Outcomes of Biology Education

The outcomes of biology education are twofold: Products and processes of science. The products of science are the **thoughts** and **attitudes** that occur as a result of learning biology. The thoughts constitute **scientific knowledge** that results from scientific investigations and explains and describe phenomena observed in the natural world. These explanations and descriptions are contained in facts, concepts, generalisations, theories, principles, and laws. These result from systematic observation, inductive and deductive reasoning and sound application of methods of scientific investigation and have been tested and shown to be reasonable, but tentatively, true.

Scientific attitudes, on the other hand, are positive feeling and dispositions about science that tend to remain through one's life. They result from having successful scientific inquiry experiences, through observations, asking questions, formulating hypotheses, testing the hypotheses, designing and conducting experiments and drawing conclusions. They enable learners to engage fruitfully in scientific investigations.

Scientific Knowledge

Scientific knowledge consists of **scientific facts, concepts, generalisation, theories, principles,** and **laws**. These building blocks of knowledge are different but related. It is necessary for the biology teacher to identify these building blocks of knowledge during the preparation of schemes of work and lesson plans. This is important because the teacher will need to use different instructional approaches depending on the nature of the building blocks. The building blocks are represented diagrammatically in Figure 3.1.

Scientific Facts

Scientific facts are the fundamental attributes of the system of ideas in science. A fact is a known single piece of information that underlies the clarity of a given scientific phenomenon, event or process. It is observable, true and usually concrete. There is no theory behind a fact. The following examples illustrate the nature of scientific facts:

- The earth rotates on its axis once about every 24 hours;
- 76 per cent of all animal species are insects; and

- Green plants contain chlorophyll in the cells of their leaves.

Scientific Concepts

A scientific concept is a mental picture of a natural phenomenon. It is a mental picture of an idea, object, phenomenon, process or event based on a class of things or attributes that are similar and which explain regularity in the observations. A concept is usually identified by its attributes – i.e. Dimensions of variation or a set of objects - the similar things that distinguish the concept. Scientific concepts are thus ideas which usually combine several facts or observations. The concept exists in the mind as an abstraction and is not to be confused with its label which is usually by a symbol. This is because scientific concepts are usually multi-ordinal- i.e. they take a variety of meaning in different contexts.

The examples given below show that a scientific concept is broader than a fact:

- **Phototropism:** Green plants bend toward the light;
- **Photosynthesis:** Green plants use light to manufacture their own food; and
- **Volume:** Occupies space, 3-dimensional, LxWxH.

Some concepts have fewer attributes and these are easier for pupils to learn. Others have for too many interrelated attributes and are difficult for them to learn.

Scientific Generalisations

Generalisations are broad ideas linking several similar concepts. In generalisations, several instances must be examined for the question that is being asked. A generalisation is characterised by two factors: it has predictive power and it has explanatory power. It predicts accurately what will happen in similar, but different situations and explains satisfactorily why it happened. For example, the idea that plants need light and water to make food is a generalisation encompassing at least three concepts: plants need food, plants need light, and plants need water. Pupils can form generalisations from their inquiries by determining the parameters that might impact on their inquiry and investigating them. Examples of scientific generalisations include the following:

- Green plants are living things;

- To maintain a species, living things must reproduce; and
- All materials on earth are formed from the 92 occurring elements.

Scientific Theories

Theories are generalisations that appear to be true but which may change if new evidence is produced that disapproves them. While they may be tentative, theories are a carefully constructed system of logical reasoning derived from well-formed assumptions regarding the basic nature of the physical world. Theories are usually very difficult to be proved. For example, according to the atomic theory, the atom contains a dense nucleus made of protons (with positive electric charges) and neutrons (with no electric charges) and a vast external space filled by rapidly moving electrons (with negative charges and almost no mass). The theory is usually represented by the **atomic model** (Fig. 3.2) that resembles the solar system model.

The atomic theory is based on indirect evidence since no one has actually seen an atom. But if it is true that atoms are constructed in this way (atomic theory) they should behave in certain ways when undergoing chemical or nuclear reactions. Many experiments have shown that they do behave as expected. Even though no one can say they have seen an atom or verify its structure, the atomic theory tells us what an atom ought to look like.

Other examples include the following:

- **Cellular Theory of Life**: Living things are made of cells;
- **Theory of Evolution**: Species adapt to their environments, and those that are most fit survive;
- **The Germ Theory of Disease**: Infectious diseases are caused by microorganisms that are capable of being transmitted from one individual to another; and
- **Theory of Ozone Shield**: Ozone in the upper atmosphere shields the earth's surface from harmful ultraviolet rays by absorbing them.

Scientific Laws

Scientific laws are theories that have been proved beyond reasonable doubt to be true. They are more general than theories and are therefore relatively fewer than theories. This is because laws imply absolute truths yet science shuns authority and maintains a sceptical attitude. Examples of scientific laws are given below:

- **The Law of Segregation**: During reproduction, the two factors

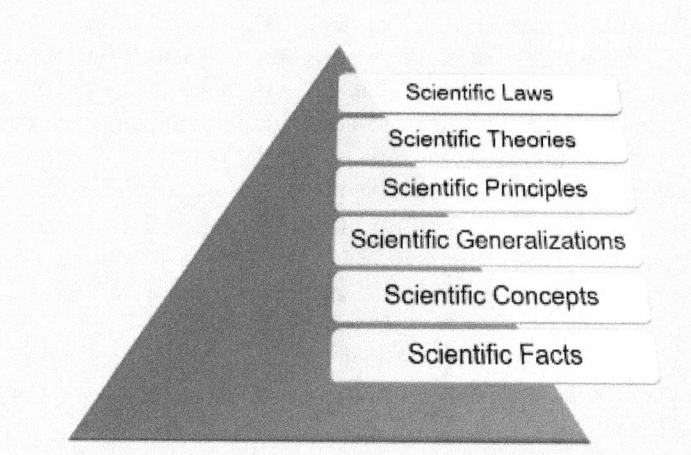

Figure 3.1: A quantitative representation of the building blocks of the structure of science

  that control each trait separate (segregate), with one factor from each pair passed to the offspring;
- **The Law of Independent Assortment**: It describes how different genes independently separate from one another when reproductive cells develop; and
- **Biogenetic Law**: All living things come from pre-existing living things.

Scientific Principles

Scientific principles are factual statements predicting interrelationships among concepts for which almost enough evidence exists but limitations also exist which we do not know. A principle can be applied to wider systems than a law (Twoli, 2006). One well-known example of a scientific principle is the **Archimedes' Principle**: *A body immersed in water displaces* *water equivalent to its volume.*

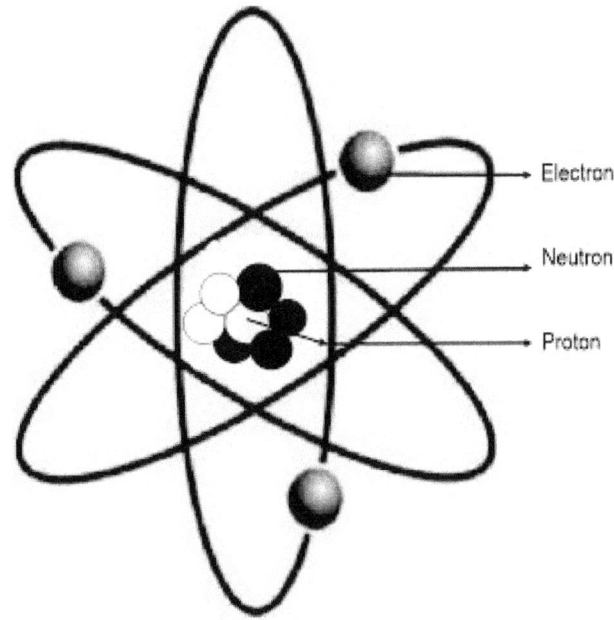

Figure 3.2: The Atomic Theory model

Scientific Attitudes

Scientific attitudes are positive feelings and dispositions about science that tend to remain through one's life. They result from engaging the learners in successful scientific inquiry experiences and processes including observations, asking questions, formulating hypotheses, testing the hypotheses, designing and conducting experiments and drawing conclusions. They also enable learners to engage fruitfully in scientific investigations.

Examples:

- Open-mindedness;
- Curiosity;
- Parsimony;
- Flexibility;
- Creativity and innovativeness;

- Intellectual honesty;
- Demonstrating resourcefulness, relevant technical skills and scientific thinking necessary for economic development;
- Seeing science as an important part of our culture;
- Understanding science as the creation of cooperative work;
- Seeing science as a critical factor for the improvement of human welfare as well as technical development;
- Seeing science as a process as well as a product;
- Awareness of social and national development;
- Sensitivity to changes and new problems;
- Willingness to take responsibility for one's judgment and ideas;
- Respect for other people's judgment and ideas;
- Freedom from superstition;
- Critical thought;
- Positive self-confidence in the judgment of ideas with respect to science and science learning;
- Patience;
- Tolerance for differences and disagreements from others;
- Listening to others;
- Self-reliance;
- Independence of thought and action;
- Judgment on the basis of sound scientific evidence and facts (objectivity);
- Faith in cause and effect relationship;
- Planned procedure in solving problems;
- Interest in biological phenomena and practical skills;
- Love for nature as well as human beings; and
- Willingness to suspend judgment until the availability of evidence.

Processes of Science

Science is the process of obtaining knowledge; scientists do science through the processes of science (the scientific skills). Doing science means applying the processes of science. Scientists do science through careful and appropriate application of the scientific processes to questions asked as a result of wondering about natural phenomena. Science process skills are abilities which are applied in understanding scientific phenomena through raising and answering questions, developing a theory or generalisation and discovering new information about the phenomenon (knowledge creation).

There are two major types of scientific skills, namely, basic process skills and integrated process skills.

Basic Process Skills:

- Raising questions;
- Observing;
- Comparing/Finding patterns and regularities;
- Classifying;
- Communicating;
- Manipulating;
- Measuring;
- Quantifying;
- Predicting; and
- Inferring.

Integrated Process Skills

- Controlling variables;
- Experimenting;
- Hypothesising;
- Defining operationally;
- Interpreting data; and
- Modelling.

The various science process skills and their specific indicators are summarised in Table 3.2.

Teachers are encouraged to ensure that students learn biology the way scientists do science. The students should apply processes of science to find out how scientists think and work and to investigate their own questions in a manner similar to the way scientists conduct their inquiries. They should use scientific skills to construct knowledge by asking questions, making observations, taking measurements, collecting data, organising and interpreting data, predicting the outcomes of manipulating one variable while keeping the others constant, formulating and testing hypotheses, developing experiments, inferring reasons for what they observe and communicating their models to others. These activities lead pupils to the facts, generalisations, principles, theories, and laws which scientists have established (Matins, 2000).

For the students to apply their scientific skills, they first have to develop the skills. Scientific facts, generalisations, principles, theories, and laws are springboards for students to use as they explore and develop the skills. The skills are thus developed using the content of science as the vehicle. For example, the skill of observing can be learned using plants, animals, rocks and moving objects. Students can learn to classify things using leaves, pictures of plants, rocks, and soil. In the process, they measure items they use in their experiments and make inferences from the information they record. Thus, from the use of scientific skills to inquire into science, pupils learn and develop the skills that help them think and reason while they are discovering scientific knowledge. They do science the way scientists do science and they begin to acquire scientific literacy. It is thus more important for pupils to learn the process skills than to learn the content alone. They should learn to do science than reading about science. The science process skills form the core of inquiry-based, hands-on science learning.

It is important that students develop basic process skills before learning integrated skills. The latter require deeper levels of thought or cognition than do basic skills. For example, pupils in lower primary classes cannot devise complex classification systems such as construction and use of a dichotomous key, nor control variables, formulate hypotheses, define operationally, plan and execute experiments and develop models. Many pupils will, however, work with these process skills if they are adequately guided by their teachers and working in groups. A good foundation of science in primary schools, therefore, predisposes the necessary scientific skills for the learning of biology in secondary schools.

Table 3.2: The Indicators of Scientific Skills

Science Process Skill	Key Indicators of Mastery of the Skill
Raising Questions	• Asking the what, how and why questions that may not necessarily lead to an investigation • Asking action questions that require investigation/inquiry • Asking hypotheses driven questions • Putting questions into a form which indicates the investigation that has to be conducted
Observing	• Identifying objects by using one or more of the senses or aids to the senses for the study of details • Describing objects or events by counting, comparing, estimating, and measuring • Discerning the order in which events take place • Noticing relevant details of the object and its surroundings • Describing properties of objects accurately such as shape, colour, size, and texture • Identifying similarities and differences • Describing changes in objects

Classifying	• Identifying key properties by which objects can be sorted
• Identifying properties similar to all objects in a collection
• Sorting or ordering objects or ideas accurately into groups or categories based on their properties
• Forming subgroups
• Establishing own sorting criteria
• Providing sound rationale for classification
• Developing and using complex classification systems
Manipulating	• Using the five senses to describe objects
• Handling and manipulating tools and materials with care for safety and effectiveness
• Assembling parts successfully according to a plan
• Working with the degree of precision appropriate to the task in hand
Finding patterns and relationships/Comparing	• Putting various pieces of information together and inferring something from them
• Finding regularities or trends in information, measurements or observations
• Identifying an association between one variable and another
• Realising the difference between one variable and another
• Checking an inferred association or relationship against evidence
Communicating	• Describing objects accurately
• Listening to others' ideas and responding to them
• Writing results of an observation of an object, or event using pictures, words, or numbers
• Talking, listening, or writing to sort out ideas or linking one idea with another or clarify ideas
• Making notes of observations in the course of an investigation
• Constructing and using written reports, diagrams, graphs, or charts to transmit information learned from investigations
• Choosing an appropriate means of communication so that it's understandable to others
• Using sources of information effectively
• Verbally asking questions, discussing, explaining or reporting observations
• Providing descriptions such that others can identify known objects
• Verbalising thinking

	•	Writing clear reports of investigations, the question asked, the experimental design used, results and conclusions drawn, using tables, graphs
Measuring	•	Determining dimensions (such as length or area), volume, mass, or weight or time, of objects or events by using instruments that measure these properties
	•	Comparing and ordering objects by length, area, volume,
	•	Measuring properties of objects or events by using standard units of measurement
	•	Measuring volume, mass, weight, temperature, area, length, and time using appropriate measuring instruments
Quantifying	•	Recording observation using numbers and correct units
	•	Recording readings correctly
	•	Changing qualitative values into numerical values
	•	Calculating proportions, percentages, ratios, volume, surface area
Hypothesising	•	Attempting to explain observations or relationships in terms of some concept or principle or conceptual relationship
	•	Applying concepts or knowledge gained in one situation to understand, explain or solve a problem in another
	•	Identifying questions or statements that can be tested
	•	Stating a problem to be investigated as a question
	•	Stating the expected outcome of an experiment/investigation
	•	Realising the tentative nature of any explanation
	•	Designing statements, questions, and predictions that can be tested by an investigation
	•	Recognising that there can be more than one possible explanation of an event or phenomenon
	•	Recognising the need to test explanations by gathering more evidence
	•	Suggesting explanations that are testable, even if unlikely

Experimenting	• Deciding what equipment, materials are needed for an investigation
• Identifying what is to change or be changed when different observations or measurements are made
• Identifying what is to be measured or compared
• Considering beforehand how the measurements, comparisons are to be used to solve the problem
• Designing an investigation to test a hypothesis
• Selecting a suitable design for an investigation to test the hypothesis
• Determining a reasonable procedure that could be followed to test an idea or hypotheses
• Following instructions for an experiment
• Developing alternative ways to investigate a problem
• Manipulating materials
• Performing experiments
• Identifying testable questions
• Designing own investigations procedures
• Formulating valid conclusions
• Recognising limitations of methods and tools used in experiments
Predicting	• Using evidence in terms of observations and prior knowledge of similar events to guess the outcome of an event
• Explicitly using patterns or relationships to make a prediction
• Justifying how a prediction was made in terms of present evidence or past experience
• Making use of evidence to formulate the sequence of the forthcoming process of action or outcome
• Using patterns or relationships to extrapolate cases where no information has been collected
• Forecasting a future event based on their prior experiences like observation, inferences, or experiments
• Showing caution in making assumptions about the general application of a pattern beyond the available evidence
• Making use of patterns to extrapolate cases where no information has been gathered
Interpreting	• Putting various pieces of information together to make a statement
• Finding patterns and regularities in observations or results of investigations
• Identifying an association between one variable

	and another
	• Ensuring that a pattern or association is checked against all the data
	• Organising and stating the information that is derived from an investigation
	• Showing caution in making assumptions about the general applicability of a conclusion
	• Constructing diagrams, graphs and charts of the data to organise and explain information correctly
Controlling variables	• Identifying the manipulated variables, responding variables, and variables held constant in an experiment
	• Controlling variables in an investigation
	• Identifying variables that can affect an experimental outcome, keeping those constant while manipulating only the independent variable
	• Identifying variables that may affect the dependent variable as stated in the problem
	• Assigning limits of control of the selected variable in an investigation
	• Proposing the degree of freedom of variables in an experiment to test the hypotheses
Inferring	• Suggesting an explanation for events based on observations
	• Making statements about an observation that provides a reasonable explanation
	• Distinguishing between an observation and an inference
	• Analysing cause and effect of decisions
Defining operationally	• Stating definitions of objects and events in terms of what the object is doing or what is occurring in an event
	• Stating how to measure a variable in an investigation
	• Defining the variable according to the actions or operations to be performed on or with it
	• Formulating a meaningful statement that generates a sense of understanding
	• Stating definitions of objects or events based on observable characteristics
Modelling	• Using pictorial, written, or physical representation to explain an idea, event, or object

> **Box 3.1**
>
> **Discussion Question**
>
> 1) Why is it necessary to discern the building blocks of knowledge before you teach biology in every lesson?

Conclusion

Biology being a science, teachers are required to ensure that students learn biology the way scientists do science. The students are encouraged to develop and apply scientific skills in investigating their own questions in a manner similar to the way scientists conduct their inquiries. In this way, the biology students construct knowledge by asking questions, making observations, taking measurements, collecting data, organising and interpreting data, predicting the outcomes of manipulating one variable while keeping the others constant, formulating and testing hypotheses, designing and conducting experiments, inferring reasons for what they observe and communicating their findings to others. These activities lead pupils to the know and understand the building blocks of knowledge such as facts, generalisations, principles, theories, and laws which have been established by scientists. In addition, the students are able to develop the scientific attitudes that are enduring and that can help them adopt good habits in problem-solving in their everyday lives.

References

Gardner, P.L. (1975). *The Structure of Science Education* London: Murray
    Jenkins, E. and Whitfield, R. (Eds) (1974) *Readings in Science Education: A sourcebook* London: McGraw Hill.
McComas, W. F. (1998) *The Nature of Science in Science Education: Rationales and strategies* London: Kluwer Academic Publishers
Martin, David J. (2000) *Elementary Science Methods* 2nd Edn. Belmont: Wardsworth/Thomson Learning
Mayr, E (2000) *The Growth of Biological Thought* 11th Edn. Cambridge: The Belknap Press.
Medawar, P.B (1969) "Two conceptions of Science" in *The Art of the Soluble*. London: Penguin.
Phenix, P (1964) *Realms of Meaning: A Philosophy of the Curriculum for General*

Education.

Robinson, J.T (1968) *The Nature of Science and Science Teaching.* Belmont: Wordsworth. Shodhganga.inflibnet.ac.in>bitstream(downloaded on 10/10/2018

Twoli, N. W. (2006). *Teaching Secondary School Chemistry: A textbook for teachers In developing countries.* Nairobi: Nehema Publishers

THE STUDY of Biology in secondary schools in Kenya aims at providing the learners with the requisite biological knowledge as well as scientific skills

Photo: GLOBAL YOUTH GROOVE/FLICKR

CHAPTER FOUR

THE SECONDARY SCHOOL BIOLOGY CURRICULUM

◆ The Goals of Teaching Biology in Secondary Schools ◆ Objectives of Teaching Biology in Secondary Schools ◆ Content of the Biology Curriculum in Schools ◆ Organisation of Secondary School Biology Curriculum ◆ Development and Progress of Secondary School Biology Curriculum ◆ Factors Influencing Secondary School Biology Curriculum Changes

Introduction

Curricula changes are the order of the day in many developed and developing countries alike. In Kenya, the 8-4-4 curriculum was introduced in 1986. The new biology curriculum was published in two different syllabuses, namely, biology and biological sciences. However, the two were later merged into the current biology curriculum. The study of biology in secondary schools in Kenya aims at providing the learners with the requisite biological knowledge as well as scientific skills and attitudes which should enable them to transform the environment for their own benefit and that of families and communities. The curriculum is broad-based in nature and anchored on biological concepts and principles as well as scientific skills and attitudes.

This chapter outlines the goals, objectives, content, and organisation of the biology curriculum in Kenya and its historical development and progress.

The Goals of Teaching Biology in Secondary Schools

The goals of teaching biology in the secondary schools of Kenya include the following:

- To help the learners recognise the nature and scope of biology;
- To develop in the learners the themes, concepts, theories, and laws of biology, think critically about them and apply them in conducting investigations;
- To develop in the learners inquiring minds through scientific approach;
- To develop in the learners the scientific attitudes needed in transformative learning and living in the society; and
- To develop in the learners the scientific skills needed for solving familiar and unfamiliar problems in biology.

Objectives of Teaching Biology in Secondary Schools

The objectives of teaching biology in secondary schools are to enable learners to:

i. Develop a better understanding of the nature of biology in terms of:

- The scientific process;
- Interrelationship and interdependence of different branches of science, especially an awareness of the place of biology among other school science subjects and in society at large; and
- Role of the scientific process in an inquiry into the living matter.

ii. Develop an understanding of biological phenomena, in terms of:

- **The interrelationship between plants and animals and between man and environment;**
- Ability to observe and identify features of familiar and unfamiliar organisms and the functions of the whole organism or its parts;
- **Application of relevant biological knowledge to improve human health: individuals, family, community;**
- **Application of relevant biological and understanding to social and economic conditions in rural and urban settings; and**
- Making use of locally available materials and information to improve the quality of life.

iii. Acquire scientific skills necessary in solving practical problems, including the following:

- **Demonstrating resourcefulness, relevant technical skills and scientific thinking necessary for economic development;**
- **Experimentation: Planning, designing, handling and setting up apparatus and equipment, conducting, and evaluating experiments and projects to understand biological phenomena;**
- Construction: making improvised aids, making minor repairs, etc;
- Drawing: drawing diagrams of experiments done and specimens used;
- **Accurate observation: identifying features of familiar and unfamiliar organisms record the observations and make deductions about the functions of parts of organisms;**
- **Communicating biological information in a precise, clear and logical manner;**
- Problem-solving skills: Identifying a problem, collecting relevant evidence, organising and interpreting data, formulating a hypothesis, testing hypothesis, drawing inferences and conclusions;

- Devising an appropriate scheme for solving practical problems;
- Handling and classifying information and quantitative results; and
- Recording information appropriately.

iv. Develop scientific attitudes, including:

- **Positive attitudes and interest in biology and relevant practical skills;**
- Open-mindedness;
- Curiosity;
- Tolerance/patience;
- Honesty;
- Respect for others viewpoints;
- Critical observation and thought;
- Freedom from superstition;
- Withholding judgment for evidence;
- **Cooperation: create awareness of the value of solving problems;**
- Self-reliance;
- Self-confidence;
- Independence of thought;
- Faith in cause and effect relationship;
- Planned procedure in solving problems;
- Interest in biological phenomena; and
- Taking responsibility for actions

v. Develop suitable career interests:

- **Acquire a firm foundation of relevant knowledge, skills, and attitudes for further education and for training in related scientific fields like medicine, agriculture, sericulture, teaching, etc.**

Note: The objectives in BOLD are also the objectives of teaching biology in Kenya secondary schools. It is imperative that the objectives of teaching biology in the secondary schools of Kenya are not as detailed as required, especially in terms of scientific skills and the philosophy of science.

The Content of the Biology Curriculum in Secondary Schools

The content of the biology curriculum depends on the needs of every country. In Kenya, five general themes are identified which serve to both unify and explain biology as a science:

- The diversity of life: Taxonomy, structure, and function, cooperation;
- Patterns of the perpetuation of life and development: homeostasis, reproduction, growth and development, inheritance, physiology, nutrition movement;
- Patterns of change Evolution;
- Interaction with environment: Flow of energy (Ecology); and
- Modern applications in biology (e.g. biotechnology, genetic engineering).

The secondary school biology curriculum in Kenya covers all these themes as indicated in the summary of the syllabus (Table 4.1).

Table 4.1: Summary of Content of the Secondary School Biology Curriculum in Kenya

FORM 1	FORM 2	FORM 3	FORM 4
1. Introduction to Biology	1. Transport in Plants and animals - Absorption - Transpiration - Translocation - Circulation system - Immune responses	1. Classification II - General characteristics of kingdoms - Main characteristics of major divisions of *plantae* - Main characteristics of *phyla Arthropoda* & *chordate* - Dichotomous key	Genetics - Concepts of genetics - 1st law of heredity - Sex determination - Linkage - Mutation
2. Classification I Major units of classification Binomial nomenclature	2. Gaseous Exchange - Gaseous exchange in plants - Gaseous exchange in animals - Respiratory diseases	2. Ecology - Concepts of ecology - Ecosystems - Energy flow in ecosystems - Population estimates - Adaptations to habitats - Effects of pollution on	2. Evolution - Origin of life - Evidence of organic evolution - Mechanisms of Evolution

		living things - Human diseases related to environmental health	
3. The Cell. Structure and function as seen under light & electronic microscope Organisational structure	3. Respiration - Tissue respiration - Application of anaerobic respiration	3. Reproduction in Plants and Animals - Asexual reproduction - Sexual reproduction - Chromosomes, mitosis & meiosis STDs, including HIV&AIDS	3. Reception, Response and Coordination - Reception, response & coordination in plants - Reception, response & coordination in animals - Role of hormones in coordination in mammals - Effects of drug abuse in human health - Structure/function of human eye/ear
4. Cell Physiology Membranes - Diffusion, osmosis, active transport - Water relations in plant/animal cells	4. Excretion and Homeostasis - Excretion in plants - Excretion & homeostasis in animals - Kidney diseases - Role of skin in thermoregulation and salt/water balance - Role of liver - Diseases of liver	4. Growth and Development - Growth and development in plants and animals	4. Support & movement in plants and animals - The necessity for support and movement in plants and animals - Locomotion in fish - Skeleton and muscle functioning in humans
5. Nutrition in Plants and Animals - Autotrophism - Heterotrophism - Chemicals of life (no details of chemical structure)			

The Organisation of Secondary School Biology Curriculum

The biology curriculum in Kenya is organised using the subject-centred approach. The approach mainly focuses on the acquisition of biological knowledge. The syllabus is based on the acquisition of facts and concepts by learners followed by practical work which is an integral part of the approach. The syllabus is organised in terms of the five themes discussed earlier. There is no evidence of a spiral arrangement of these themes in the syllabus. Many other countries of the world use this approach as well. The science curricular projects like PSSC, BSCS, CHEM-S, SSP, and Nuffield Science Project were all developed using this approach.

Development and Progress of Secondary School Biology Curriculum

The biology curricula throughout the world have been changing depending on the changing perspectives of education and the role of biology in transforming human lives in those countries. The current biology curriculum for secondary schools in Kenya has experienced fundamental changes since independence in 1963 (see Table 4.2) as are curricula for chemistry, physics, and mathematics. The changes have been in terms of content, teaching approaches, instructional materials, and assessment procedures. These changes are best understood by tracing them through four phases: **Experimental/latent; adoption; adaptation** and **ownership** or **indigenisation**.

Table 4.2: Phases in the Science and Mathematics Curriculum Change in Kenya since 1963

PHASE	PRIMARY SCIENCE AND MATHEMATICS	SECONDARY SCIENCE AND MATHEMATICS
Experimental/Latent	1963-1975	1963-1967
Adoption	1976-1984	1968-1980
Adaptation	1985-2002	1981-2002
Ownership/Indigenisation	2003	2003

Experimental Phase

During this phase, the traditional curriculum was in operation. However, in the wake of the demands for the fast developing technology, an experimental curriculum was also developed and piloted during the period. The information that follows indicates the long journey in the transformation of the curriculum.

a) The Traditional Biology Curriculum

Content: The traditional biology curriculum focused on the transmission of knowledge at the expense of the development of essential scientific skills and attitudes. The **content** of the traditional biology curriculum for secondary schools between 1963-1967 was developed and examined by the Cambridge University Examination Syndicate in collaboration with the East

African Examination Council up to 1973. The content had the following characteristics:

i. Basic biological concepts for the Ordinary (O) level curriculum were all introduced in Forms I and II (Junior Secondary School). The concepts focused on living things and how they functioned but excluded genetics and ecology. The concepts were further revisited in more detail in Forms III and IV (Senior Secondary School). Thus a spiral approach was used to organise the curriculum;

ii. The content for the Advanced (A) level (Forms V and VI) curriculum was more elaborate and included a detailed analysis of ecology and a molecular approach to physiology, and genetics;

iii. **Teaching:** The teaching of biology was mainly didactic whereby memorisation was encouraged at the expense inquiry. However, experiments were performed only to prove or confirm the observations and facts covered during theory classes. The curriculum, therefore, focused more on the transmission of knowledge at the expense of development of essential scientific skills and attitudes;

iv. **Teaching materials:** The instructional materials included basic textbooks written by authors mainly from Britain who looked at biological concepts from their own cultural perspectives. There were no teachers' guides although the inspectorate section of the ministry of education occasionally issued circulars addressing specific instructional issues in the curriculum; and

v. **Assessment**: The level of student achievement was **assessed** by the Cambridge University Examinations Syndicate in collaboration with the East African Examinations Council up to 1973. The latter took over the functions in 1974 and worked up to 1977. The examination items covered both theory and practical work. The items in the theory paper were of the structured and essay type. Practical examinations focused on observation, measurement, drawing and inference skills as well as applications of knowledge in new situations.

b) The Experimental Curriculum Initiatives

The experimental curriculum was developed beginning in 1965 and was implemented after 1967. Its main goal was to eventually replace the then existing traditional curriculum. The initial efforts were made by the Education Development Centre (EDC) in the United States of America (USA). At a workshop in the United States in 1961, Professor Jerrold Zacharias of the Massachusetts Institute of Technology (MIT) along with several other scholars created the African Education Program (AEP) under the auspices of EDC, funded by USAID. The purpose of AEP was to enhance the provision of knowledge, skills, and attitudes necessary for the use of technology for the development of African countries.

The AEP initiated several projects among them the African Mathematics Programme (AMP), African Primary Science Programme (APSP) and Science Education Programme for Africa (SEPA) which replaced the APSP in 1969. The programmes initially focused on primary science and mathematics and drew heavily from the Elementary Science Study Programme (ESSP) in the USA. The programme later expanded to include secondary school science and mathematics, drawing heavily from the famous programmes in biology, chemistry, and physics, namely: Biological Science Curriculum Study (BSCS); Physical Science Study Committee (PSSC); and Chemical Education Material Study (CHEM-S).

At the second workshop held in Entebbe in 1962, the participants drawn from various parts of the world resolved to develop curricula in science and mathematics for African countries both at primary and secondary schools based on the inquiry approach adopted by ESSP, BSCS, PSSC, and CHEM-S Programmes. From 1965 to 1967 several expatriate teachers and education officers in East Africa (mainly British) held a series of workshops under the umbrella of the **School Science Project** (SSP) and developed science curricula for secondary schools at the Ordinary level (Forms I-IV). The SSP was based on the philosophy of the **Nuffield Science Project** in Britain. However, the project was similar to the PSSC, BSCS and CHEM-S projects in terms of philosophy and objectives. The expatriates developed curricula and trial materials in the following areas: SSP Biology; SSP Chemistry; SSP Physics and School Mathematics Project (SMP) Mathematics. The materials were tried in some schools in East Africa and were refined for implementation beginning in 1968.

Adoption Phase

The SSP Biology curriculum was finally implemented from 1968-1981 in over 50 schools in Kenya, particularly in the fairly well equipped provincial and national schools. Many schools declined to accept the programme, citing unpreparedness of the teachers and lack of appropriate apparatus and equipment. The real problem could have been due to the fact that many Kenyan teachers were not involved in the initial development of the programme apart from not being inducted in implementing it. Within the same period (1968-1981) a majority of the schools continued to offer the traditional biology curriculum at both the 'O' and the 'A' Levels. However, both the SSP Biology and the traditional curricula were a case of **adoption** because the content, objectives, philosophy of the programmes, instructional procedures, and assessment techniques were heavily drawn from the Nuffield Biology and Biological Science Curriculum Study and the mainstream British (colonial/ traditional) curriculum respectively.

The **objectives of SSP Biology** as derived from the BSCS and Nuffield project were to:

- Make biology teaching stimulating to the pupils so that they would take up further study in biology;
- Teach biology through inquiry or discovery approach. This would enable pupils to develop the scientific skills and attitudes necessary in their day to day solution of problems;
- Enable students to develop biological concepts through the scientific inquiry;
- Teach biology based on students' experiences, interests, and environment;
- Enable students to use biological knowledge rather than simply memorise it; and
- Make biology more relevant to the needs of the learners and to their local environments.

Content: The content of the SSP Biology ('O' Level) curriculum was a radical departure from the traditional one. It included new topics like taxonomy, ecology, and a molecular approach to elementary physiology and genetics which were hitherto preserved for the Advanced level students. The content was also organised using the spiral curriculum approach.

Teaching: Teaching preferred for SSP Biology was supposed to be inductive using the inquiry approach. The focus was to be on the development of scientific skills as well as scientific knowledge. Science process skills such as observation, measuring, experimenting, recording, inferring, hypothesizing and drawing conclusions were key in the learning of SSP Biology. This approach placed heavy demands on the teachers in terms of preparation and supervision of pupils.

Teaching Materials: The teaching materials included the use of teachers' guides and pupils' manuals, textbooks and supplementary readers centrally prepared by the Ministry of Education. The materials were of high quality and specially developed for the purpose of enhancing inquiry learning. This was done to standardise the teaching process, given the radical changes involved.

Assessment: The assessment of students' learning in SSP Biology included:

- Recording of the field and experimental data and writing of comprehensive reports in notebooks which were then assessed by the teachers;
- Ministry of Education officials' verification of the students' records of the field and experimental data and reports in notebooks completed by the students every term and assessed by teachers; and
- Sitting of external examinations set and administered by University of Cambridge Examination Syndicate in collaboration with East African Examinations Council. The theory examination included essays and structured questions covering knowledge of biological concepts, experiments performed during the four years of school, and the application of results obtained in such experiments. There was no practical examination since this was covered in the school practical programme.

Adaptation Phase

In 1981, the New Mathematics Programme was scrapped by the Presidential Decree. The ministry of education in its wisdom moved swiftly to also scrap SSP Biology, SSP Chemistry, and SSP Physics. The reasons cited for the banning of the SSP Biology curriculum were:

- The teachers found it difficult to follow the suggestions in the students and teachers' books for conducting the practical;
- It focused more on the scientific process and less on providing important biological information;
- The curriculum required use of sophisticated apparatus and equipment which were lacking in most schools;
- It required too much time on the part of the students in terms of conducting practical, keeping field and experimental records and writing reports;
- It placed high cognitive demands on the students;
- It required a lot of time on the part of the teachers in terms of preparation and supervision of students' learning;
- Teachers (especially Africans) were not in-serviced in the philosophy, objectives and inquiry teaching approach of the SSP Biology curriculum;
- The students who went through the curriculum fell back to the traditional curriculum in the 'A' level course and faced many learning difficulties; and

- The programme had continued to be modified and tended towards the traditional biology curriculum.

In 1982, the SSP Biology curriculum was replaced with the Integrated Biology curriculum which incorporated aspects of the SSP Biology into the traditional biology curriculum. Students were taught the new integrated curriculum. However, from 1981-1984, a new biology curriculum was developed under the auspices of the 8-4-4 curriculum. The curriculum, which was implemented in 1986, had two syllabuses: Biology and Biological Sciences. The new 8-4-4 biology curriculum lasted up to 2002.

The **content** of the new curriculum was an integration of SSP Biology syllabus, the traditional biology syllabus and some aspects of the Advanced level biology syllabus, which was discontinued in the 8-4-4 system of education. The new curriculum included more detailed analyses of topics such as a survey of plant and animal kingdoms, chemicals of life, cell physiology, transport in plants and animals, homeostasis, genetics, evolution, and ecology than before and included many ideas previously covered by the Advanced level biology curriculum.

Teaching was also supposed to be based on inquiry approach, but many teachers preferred to continue using didactic approach due to such factors as lack of proper training; apparatus and equipment; commitment to professional delivery in the classroom; and support from quality and standards section of the Ministry of Education.

Teaching materials included pupils' and teachers' guides developed centrally at the Kenya Institute of Education (KIE) that gave clues to the teachers on objectives, duration, instructional procedures and teaching materials for each topic.

Assessment of students' learning involved continuous assessment tests (CATs) conducted regularly by the teachers in schools and external examinations at the end of the course after four years at school. Examinations are set and administered by the Kenya National Examinations Council (KNEC). Theory examinations include a written essay and structured questions, while the practical examination still focuses on the application of scientific skills and knowledge in new situations. However, there is a great need to improve on continuous assessment tests with a view to focusing not only on biological knowledge but also on scientific skills and attitudes gained.

Ownership or Indigenisation Phase

In 2003, a new biology syllabus was launched in schools. The biological science syllabus was discontinued, but some of its content was integrated into the new syllabus. The content of the new syllabus is essentially similar to the biology and biological sciences syllabuses, except that it now includes aspects of human biology and first aid that are incorporated from the latter. More significantly, the syllabus does not require a detailed analysis of chemicals of life, mechanisms of inheritance, physiological processes, taxonomy, and ecology. The move helps to place a lighter load on both the teachers and students and may be friendlier to students in terms of cognitive demand. But whether we can say we own the curriculum is a matter of debate. The underlying question to this debate is this: 'How relevant is biology education to pupils and the society at large? Does the curriculum heavily draw from society? If it does, are process skills at the centre of the learning process rather than content?'

Factors influencing Secondary School Biology Curriculum Changes

Many factors influence curriculum change in many countries of the world. In Kenya, the factors influencing the biology curriculum change in each of the four phases of curriculum change can be categorised into **External** and **Internal Pressures**.

External Pressures

The agents of curriculum development based in the USA were responding to the launching of the **Sputnik** by the then Soviet Union in 1957 saw a need to reform science and mathematics education in their countries, an idea that quickly gained currency in Britain and the rest of Western Europe. There then followed an intensive interest and massive support for subject-based curriculum change in science and mathematics. The changes that were proposed and implemented in the USA and Britain easily found their way to Africa and Kenya since they were justified on the grounds of introducing modernity. The changes were also accompanied by financial aid.

According to Hawes (1979) the impetus for change was provided by the need to help the allies of the USA and Britain to move with the times as the changes would address:

- Gaps between research findings by university researchers and what was taught in schools; and
- Demands of computer-based technology and realities of a curriculum designed in the 19th century to serve the less technical nations.

Internal Pressures

The newly independent Kenya, as part of its political aspiration, wanted to review the curricula to reflect the needs of the nation. The Ominde Commission of 1964 recognised the role of science and mathematics in the promotion of socio-economic development of the country and therefore recommended that the teaching of science and mathematics should be invigorated and given priority over other fields of knowledge. These political pressures came at a time when external pressures to reform the teaching of science and mathematics were mounting and augmented curriculum reforms. The government therefore quickly supported the changes suggested by the external curriculum initiatives. Other internal pressures that have provided the incentive for curriculum change included the Koech report (1997), Master plan for education, and The Kenya Vision 2030, among others.

Conclusion

The chapter provides the basis for planning and teaching biology in secondary schools. It is clear that biology in secondary schools in Kenya aims at providing the learners with the requisite biological knowledge as well as scientific skills and attitudes which should enable them to transform the environment for their own benefit and that of the family and community. The curriculum is broad-based in nature and anchored on biological concepts and principles as well as scientific skills and attitudes. The curriculum has undergone changes over the years in an effort to make it more relevant and to respond to the needs of the students and the nation at large. Whereas the curriculum heavily borrows from curricula in other countries, efforts have been made to indigenise it.

Reference

Hawes, H. (1979) *Curriculum and Reality in African Countries*. London: Longmans

LEARNING IS a complex process, several theories have been advanced to explain how individuals learn

Illustration: FUNDERSTANDING/FLICKR

CHAPTER FIVE
LEARNING BIOLOGY: THE THEORETICAL FOUNDATIONS

◆ The Concept of Learning ◆ How Learning Occurs ◆ Some Learning Theories Relevant To Learning of Biology ◆ Behavioural Learning Theory ◆ Cognitive Learning Theory ◆ Overall Implications of Learning Theories for Biology Teaching

Introduction

Although science (including Biology) has been taught for a long time, relatively little has hitherto been known about the psychological processes by which students learn such fundamentally different things as scientific concepts, principles, attitudes, skills, and facts. Research on these learning processes continues and the results are having some effect on the approaches to the teaching of biology. This chapter reviews known facts and theories pertaining to learning and their influence on the teaching of biology.

The Concept of Learning

Crow and Crow (1963) define learning as "A change in the individual due to the interaction of that individual and his/her environment, which fills a need and makes him/her more capable of dealing adequately with his/her environment." Learning is therefore seen as a process by which one modifies one's behaviour as a result of being exposed to a certain situation or stimulus or a series of stimuli. It is the process by which we acquire and retain knowledge, understandings, skills, attitudes, skills, and capabilities that cannot be attributed to inherited behaviour patterns or physical growth (Farrant, 2002). The process is believed to be physiological and takes place in the brain. The rate of learning depends on inherited and environmental factors.

The modified behaviour is observed or detected when the learner performs some task related to the situation or stimulus exposed to. The modified behaviour may also be exhibited in the form of attitude, interest or value. **We can, therefore, recognise that learning has oc-curred when we note a behavioural change in the learner and when we note the persistence of this change.**

How Learning Occurs: The Learning Theories

For many years learning has been a subject of research by psychologists, but remarkably, not much has resulted in improved teaching. Because learning is a complex process, several theories have been advanced to explain how individuals learn. A better understanding of the process started to emerge in the past 20 years when research into the ways students go about their learning began.

A **learning theory** is a statement about the conditions under which learning will take place. Theories of learning make it easy for us to depict how learning takes place and the conditions that facilitate learning. It is a testable prediction. A learning theory gives us the following information:

- Nature of the learning process;
- Conditions under which learning takes place;
- Clues about human thought; and
- Direction on teaching under various conditions

Knowledge of learning theories helps teachers in terms of:

- Identifying possible outcomes of learning (objectives);
- Choice of learning experiences;
- Choice of instructional strategies;
- Assessment of what has been learned; and
- Choice of teaching/learning materials.

Some Learning Theories Relevant to Learning of Biology

While the **Behaviourist Theory** was the first learning theory, the other two theories of learning focusing on student activity have come up in the last 20 years, namely, **Cognitivist** and **Constructivist Theories**. Other theories have appeared more recently building on the three, including **Humanism** and **Reconstructionist Theory** (Figure 5.1).

Behavioural Learning Theory

The behavioural learning theory is associationist in nature and considers learning as the acquisition of observable behaviours instead of internal attributes such as thinking and emotions. It is based on conditioning as the basic unit of learning. Conditioning through a series of Stimulus-Response (S-R) situations leads to the acquisition of behaviour. Reinforcement is an important process in the acquisition of behaviour.

Behaviourism was developed by psychologists like Ivan Pavlov (classical conditioning), B.F Skinner and E.L Thorndike (Operant or instrumental conditioning) and Albert Bandura (social learning theory). Many of the basic principles of behavioural learning theories were first clearly stated by Ivan Pavlov after he carried out his classic experiments of conditioned responses.

Classical Conditioning

Classical conditioning is learning in which a neutral stimulus becomes able to elicit a previously unrelated response; it involves associating a new stimulus to an innate reflex. Pavlov rang a bell to get the dog to salivate, thus demonstrating that dogs learn to make a conditioned response (CR) to a conditioned stimulus (CS) in the same way they respond to an Unconditioned Stimulus (UCS). Wherever a bell rings along with food, **reinforcement** takes place. But several abortive soundings of the bell will lead to a lessening of the quantity of saliva and ultimately **extinction** of the conditioned S-R link.

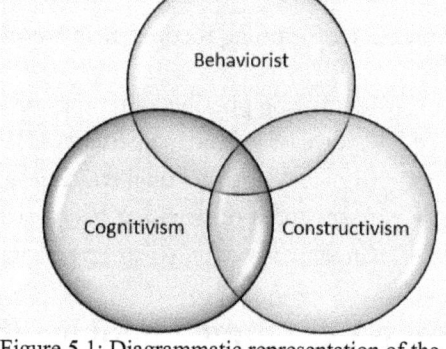

Figure 5.1: Diagrammatic representation of the relationship between the three major learning theories. (The intersection of the Venn diagram represents the more recent theories, particularly Reconstructionism)

The four principles Pavlov proposed to explain classical conditioning were, therefore:

- Reinforcement-CS-CR association can be strengthened (strengthening a response);
- Extinction–CS-CR association can be extinguished (disappearance of a response);
- Generalisation-CR can generalise in the presence of similar stimuli (similar responses); and
- Discrimination-CR can be refined through discrimination (different responses).

In terms of educational applications, classical conditioning is limited to reflex mechanisms and in turn to emotional reactions. It may account for some certain activities of pupils, such as ringing bells and timing of school activities. Others such as the effect of reward and punishment, fear of failure and pleasure from praise may be traced to conditioning.

Operant (Instrumental) Conditioning

Many other psychologists such as B.F Skinner (1904-1990) and Thorndike carried out variations of Pavlov's original experiments with animals and humans. They all realised that classical conditioning accounted for only limited types of learning.

Operant conditioning is based on Thorndike's law of effect, which states that actions followed by satisfying changes are more likely to be repeated on future occasions. An operant is a behaviour that produces consequences for the individual. Skinner's experiments on hungry rats placed in "Skinner boxes" consisting of levers, which when pressed would cause the release of food pellets, showed that rats displayed a dramatic change in behaviour by intentionally pressing the lever, often very quickly, to obtain food. This was an example of trial and error learning. The experiments show that voluntary responses are governed by the consequences that follow (positive or negative). Repeated experiments with rats and pigeons enabled B.F Skinner to develop five basic principles of operant conditioning, namely:

- **Positive reinforcement**: the desired behaviour is strengthened by presenting a positive stimulus after the behaviour occurs;
- **Negative reinforcement**: the desired behaviour is strengthened by removing a negative stimulus after the behaviour occurs;
- **Punishment**: Weakening an undesired behaviour by presenting a negative stimulus after the behaviour occurs;
- **Time-out**: Weakening an undesired behaviour by temporarily removing a positive reinforcement that supports it; and
- **Extinction**: Weakening an undesired behaviour by ignoring it.

Continuous reinforcement schedules are best for establishing new behaviours relatively quickly. Intermittent reinforcement schedules establish behaviours relatively slowly, but the resulting behaviours are more resistant to extinction. Fixed interval, variable interval, fixed ratio, and variable ratio schedules each produce characteristic patterns of responding:

- **Fixed interval**: Reinforce after regular predetermined intervals;
- **Variable interval**: Reinforce after several time intervals of different lengths;
- **Fixed ratio**: Reinforce after a predetermined number of responses; and
- **Variable ratio**: Reinforce after a different number of responses.

The applications of operant conditioning to education are as follows:

- Use of programmed instruction, which was fashionable in the 1970s. Components of any skill or set of information are structured into increasingly comprehensive units to facilitate learning. The earliest programmes were directly based on Skinner's rat and pigeon experiments;
- Knowledge of results or feedback has reinforcement value; and
- Children can be encouraged to learn using extrinsic rewards (incentives, praise, and reproof).

Albert Bandura's Social Learning Theory

According to the social learning theory, learning occurs when we observe and imitate other people. The four processes of observational learning proposed by Bandura are:

- Attention: capture learner's attention;
- Retention: demonstrate/explain to ensure learner retains and produces behaviour;
- Production: give related problems for them to solve; and
- Reinforcement: motivate/reinforce for correct answers. Motivation may be through direct reinforcement, self-reinforcement, or vicarious reinforcement (observation of a model being reinforced and increases the likelihood of imitation).

Regarding the application of social learning theory to education, the following are relevant:

- Observational learning can happen only if students can identify and attend to the model behaviour, can hold what they observe in memory, and have the necessary skills to produce the model behaviour; and
- Encouraging students to set their own goals, observe their own behaviour, and administer their own reinforcements so as to achieve self-management of learning.

Overall Educational Applications of Behavioural Theory

The following behaviour modification techniques based on the theory could be used to facilitate learning:

- **Shaping**: Involves reinforcing successive approximations of a final, desirable behaviour; this is one way in which rewards move in the desired direction. The Premack principle of requiring work to be done before the student chooses the reward is the example of shaping. The Premack principle states that any frequently occurring behaviour can serve as reinforcement for any infrequently occurring behaviour;
- **Token economies**: Tokens are awarded for desired behaviours and later cashed in for tangible reinforcers;
- **Contingency contracting**: This refers to the use of written documents to specify desired behaviours and reinforcements;
- **Time-out, extinction and response cost**: This is used to weaken undesired behaviours. Time-out works best with disruptive students. Time-out procedures are ways to reduce inappropriate behaviours by immediately denying a student all reinforcements for an affixed short period of time. Response cost refers to a way of reducing the frequency of undesired behaviours by removing a specified reinforcement whenever the behaviour occurs; and
- **Punishment**: This is used to weaken undesired behaviours, such as caning, but is often ineffective.

Cognitive Learning Theory

Among the contributors to this theory (the associated theories indicated in parenthesis) are:

- Jerome Bruner (**Meaningful learning**)
- David Ausubel (**Meaningful reception learning**)
- Robert M. Gagne (**Information processing learning theory**)
- Jean Piaget (**Developmental theory of learning**)

Cognitive theorists, among them Jerome Bruner, David Ausubel, Robert Gagne, and Jean Piaget, believe that activities do occur in an organism between stimulus and response and that these activities are crucial to an understanding of behaviour. Cognitive theorists addressed the following questions:

- What is learning?
- How does learning take place?
- What are the implications of the learning process of teaching in the classroom?

The cognitive learning theory focuses on cognitive processes by which we acquire and use knowledge. They all indicate that learning results in organised storage of information in the learner's brain and this organised complex is referred to as a **cognitive structure**. The theory, therefore, views learning as the acquisition of knowledge structures that lead to a change in the learner's meaning of experience. It is predicated on the premise that changes in knowledge occur during the learning process during which the learner actively processes incoming information or ideas in terms of their existing ideas (cognitive structure).

According to the cognitive learning theory, learning is defined as a process by which the learners modify their behaviour as a result of being exposed to certain situations. The process is physiological and takes place in the brain. The modified behaviour is observed or detected when the learner performs some task related to that situation. The modified behaviour may be exhibited in the form of performance, attitude, interest or value.

David Ausubel's Meaningful Reception Learning

According to Ausubel, meaningful reception learning occurs when new information is linked to existing relevant concepts in the learner's cognitive structure. The subsumption process can be summarised diagrammatically (Figure 5.2):

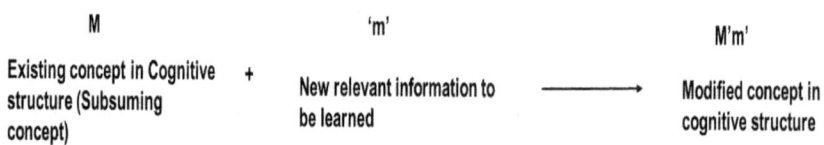

Figure 5.2: Ausubel's theory of Meaningful Reception Learning

For a period of time, the new information learned ('m') can be recalled almost in its original form, but in time, it will no longer be dissociable from the subsuming concept. Meaningful reception learning occurs when:

- New ideas provided by others are easily assimilated into knowledge schemes (cognitive structure);
- Relationships among ideas exist; and
- Material to be learned is well-organised.

Reception learning is likely to be less meaningful when relationships among ideas are ignored. It is also possible to learn new information with little or no linkage to existing elements in the cognitive structure. This is regarded as rote learning.

Meaningful reception learning is expressed in the learner's ability to solve relevant new problems. For example, if learners have meaningfully learned some aspects of gene structure, they should be able to solve new relevant problems on genetics. Problem-solving ability thus derives from cognitive structure differentiation (new information and previously learned information coalesce into a whole) and is concept specific. Ausubel suggests that properly designed instructional sequence (**advance organiser**) introduced prior to new information would facilitate learning of the information. However, it should be more general and more abstract than the information to follow and it should facilitate meaningful learning of the new material. It should be like the new information to be learned with existing concepts in cognitive structure. Advance organisers can help learners establish a meaningful learning set. For example, the necklace with beads of different colours and the holding string acting as a DNA strand may be used as a cognitive bridge prior to instruction on the structure of DNA or genes. In the same way, the concept of structure and function complementarity when used as a cognitive bridge prior to instruction on the nature of xylem elements will facilitate meaningful learning of the concept.

In terms of its application in education, Meaningful reception learning theory influences the use of well-designed lectures, textbooks and programmed learning materials.

Jerome Bruner's Meaningful Learning

According to Bruner, meaningful learning occurs when the structure of the subject matter is well organised and when learners fully participate in discovering the structure. In this case, structure refers to the interrelationships among basic ideas and how they relate to each other. For example, the use of concept maps helps in achieving meaningful learning by displaying these interrelationships in a set of simple, easy to understand diagrams that minimise the need for rote memorisation of isolated facts. Similarly, each idea can be organised from simple to more complex in a spiral manner. On this basis, Bruner advocates the use of a **spiral curriculum** in schools. In addition, each learner is encouraged to discover ways of relating new information to the existing cognitive structure in an effort to achieve meaningful learning, that is, to discover how ideas relate to each other and to existing knowledge.

Bruner's theory is applied in many important ways in education:

- **Schemes of work**: Concepts and ideas are usually clearly delineated and their sub-concepts identified when drawing out schemes of work;
- **Teaching from Simple/Unknown**: Teaching of concepts or ideas should begin from simple to more complex forms and from what is known to what is not known;
- **Spiral arrangement**: Concepts and ideas should be taught progressively in a spiral manner from simple to more complex forms as students grow intellectually over the years;
- **Discovery approach**: Discovery approach, guided by the teacher, can be used whereby emphasis is on learners discovering how ideas relate to each other and to existing knowledge; and
- **Active Participation**: In teaching, use contrast, guessing, participation and awareness to stimulate discovery.

Robert Gagne's Information Processing Model

Information processing model of learning attempts to explain how learning occurs. It focuses on the transformations which occur between the 'input' of external stimuli and the 'output' of the learner's behaviour. Information from the environment (external stimuli) in the form of events and attributes affects the receptors leading to coding. The initial perception then takes place.

The selection of the stimuli depends on the attributes, which make them noticed such as intensity, colour, shape, size, as well as the attributes of the learner including alertness and cognitive strategies (what is looked for). Cognitive strategies are mental skills that control how we process information and include rules, concepts, discriminations and associations that form a hierarchy.

The processing of information begins when the information (stimuli) enters the short-term memory (a few seconds) and is again coded. If it is to be remembered, it is transformed once more and enters the long-term

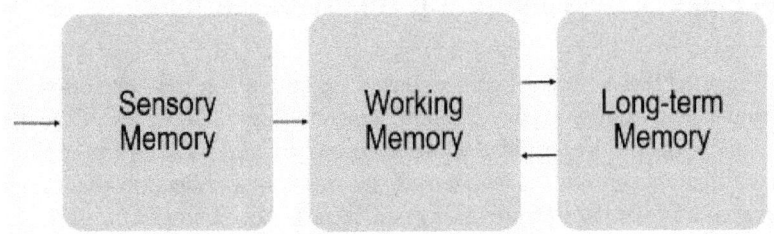

Figure 5.3: Information Processing Model

memory where it is stored for later recall (Figure 5.3). The short-term memory can hold on to new learning for as long depending on the nature of item learned (cognitive demand) and interference, which leads to forgetting. 'Forgetting' is a problem of retrieval. If the teacher overloads the learner's short-term memory store, learning becomes difficult. What is seen as less complex to the teacher may be too complex to the child; the child looks at the bits that make up the complexities and should be trained to see things in details and to unify bits into a whole. If the long-term memory store is damaged, information cannot be retrieved. To facilitate learning, the learner needs to rehearse (mental repetition).

From the short-term or long-term memory, information passes to a response generator which activates behaviour (performance). The efficiency of this process depends on the motivation of the learner (state of need such as hunger, prior knowledge, features of stimuli), personal meanings of input (stimuli), and cognitive strategies.

The information processing model has several applications in education:

- The acquisition of knowledge is a process in which every new capability builds on a new foundation established by previously learned capabilities – i.e. students are ready to learn when they have mastered the prerequisites; that's when they acquire the necessary capabilities through preceding learning;
- Knowledge is hierarchical. A learning hierarchy is a map of the knowledge that children must have at any point if they are to learn a particular new element. It is 'a network of ladders of elements of knowledge which lead up to the terminal behaviour the student is required to learn"
- Performance at lower levels of a task equips necessary skills for performance on a new and more difficult task; and
- In learning, a child first apprehends the stimulus situation, proceeds to stage of acquisition, storage, and retrieval.

Jean Piaget's Developmental Theory of Learning

Jean Piaget, a Swiss psychologist, focused his studies on the development of children's thinking patterns (Genetic epistemology). Many of his experiments have been repeated elsewhere in the world and most of his basic ideas have been accepted. Piaget viewed the mind as a collection of cognitive structures he called **schemata (sing. schema)**. According to him schemata are opened, enlarged, divided and connected in response to the influx of information into a person's mind. As information gets into the schemata, the learner strives to make sense by connecting to something they already know, connecting in ways unique to their method of thinking and prior experiences. The learner is achieving equilibration or adapting to a new situation (adaptive behaviour).

Equilibration involves two processes, namely, **assimilation** and **accommodation**. Assimilation is the taking in of information and fitting it into existing schemata (registering sensory input). However, there comes a time when either the existing schemata are too filled with information to warrant enlargement, or there are no existing schemata appropriate to accommodate certain new pieces of information. In these cases, new schemata are opened to accommodate exposure to new experiences and observations. The opening of new schemata (either by splitting off existing ones or opening brand new ones), is called **accommodation**. This accommodation is the adjustment of cognitive structures. However, humans accommodate if the information being learned is not completely new. For example, children will engage in activities while expecting certain results. If they find it is very new, they lose interest. The sequence of developing cognitive structures is believed to be common to all hu-mans.

The learning theory that emerges from Piaget's work can be summarised by saying that cognitive change and learning take place when a schema, instead of producing the expected result, leads to accommodation that establishes a new equilibrium. As a result, the human mind is a complex network of schemata, which are intricately connected to each other in patterns completely unique to the individual. According to Piaget, learning is thus an active process. To learn something new, learners must be able to connect it to something they already know. They must construct the personal meaning of the new material, either by assimilating it into existing schemata or accommodating it by forming new schemata that are connected to existing ones. In this case, learning can be inhibited by certain barriers, namely: Small accommodation structures; lack of motivation; and entry behaviour (lack of **mastery of prerequisites**).

Piaget's Stages of Cognitive Development

Piaget suggested four stages of cognitive development commencing with birth (Table 5.1):

- Sensorimotor stage;
- Preoperational stage;
- Concrete stage; and
- Formal operational stage

He believed that all people progress through the same four stages in the same unvaried sequence. However, not all children go through the stages of development at the same pace. Each stage represents a more advanced capability for cognitive processing than the previous stage. It is essential that a science teacher in general and biology teacher, in particular, understands the intellectual capabilities and limitations of children at various stages.

Table 5.1: Cognitive Development Stages and Support Needed for Learning

Stage	Characteristics	Ways to foster development
Sensorimotor (0-2 years)	- Thought is based on sensory input - Permanence of objects: a coloured object when hidden will be looked for - i.e. object 'seen' in mind. - Learns to suck to obtain food. - Holds things put in hands - Follows moving objects with eyes and turns head to look for the source of the sound.	- Opportunity to act on environment-e.g. working with puzzles with large pieces, playing with push toys, exploring the house and yard, walking unaided up and down stairs
Preoperational (2-7 years)	- Develops symbolic functioning(that provides human superiority)-e.g. learns to talk, but thinking closely tied to actions - Construction: building things from materials available - Imitation - Egocentric thought: tell themselves what to do - No formal conceptions-objects have no relationships with each other; have life and feelings - Reversibility difficult - Class inclusion not possible therefore no formal concepts	- Language and symbol use - Drawing and interpretation of pictures (colour) - Picture books and drawing materials provided - Building blocks and toys for pulling
Concrete operations (7-12 years)	- Inferential logic: A>B, B>C, A>C - Hierarchy of classes-e.g. using brown and white wooden beads - Which are more? Classifying according	- Hands-on experiences - Availability of materials children can sort, count, classify, put in order-e.g.

	to colour, shape or size. - Ordering: using physical objects e.g. length or size - Difficulty in keeping in mind more than one relationship at a time e.g. classifying according to colour/size; and time sequence - Difficulty in understanding cause and effect relationship - Depend on what they see and hear-or think they see and hear (perceptions) rather than on any other evidence presented - Groupings (combinability, reversibility, associativity, identity, tautology) - Conservation of number, length, mass	seeds, leaves, matchsticks, buttons, bottle caps, string and thread - Simple experiments on cause and effect ideas - Storytelling - Using words carefully
Formal operations (12-15 years)	- Hypothetical-deductive thought (abstract reasoning): hypotheses developed from which to work out the parts; more interested in the real world, how things work, what they are made of, where they come from, etc; sees situations from different viewpoints. - Difficulty in verbal problems, planning ahead - e.g. steps in an experiment; does not test hypothesis systematically. - Carries out family tree classification - Can carry out multiple classifications like classifying children according to height, clothing, etc. - Propositional logic-i.e. forms argument independent of concrete evidence - Combinational system —operates with several propositions at once; sees need to vary one factor at a time in test situations. - Proportionality develops –e.g. speed, density, SA/VR ratio	- Abstractions in terms of problems, hypotheses, probabilities, correlations, proportions. - Formal reasoning skills -Concrete examples be used in the discussion of abstract ideas-e.g. plastic models to illustrate chemical formulae.

Jean Piaget's revelations have been widely applied in education. For example, in curriculum organisation, **the spiral approach** is employed in which a concept is introduced in one class and developed step-by-step during the later years. This allows the concept to be examined and re-examined over and over again as the children's thinking develops, and as they become capable of understanding the more difficult aspects of the concept.

Constructivism

Constructivism has a long history in cognitive psychology and has its origin in the work of Jean Piaget, David Ausubel, Robert Gagne and Jerome Bruner among others. Cognitive theories view learning as the acquisition of knowledge structures that can lead to a change in the meaning of experience. The learner actively processes incoming information in terms of their existing ideas. David Ausubel says of this: *"...the most important single factor influencing learning is what the learner already knows"*.

Constructivists, on the other hand, argue that learning is not just the adding of new knowledge to the existing one but involves a restructuring of knowledge in the mental system (cognitive structure) by the learner. They believe that learners must construct meaning for themselves; that the only learning that can take place is that which is connected to the individual's already existing knowledge, experiences, or conceptualisations. What children learn is not a copy of what they observe in their surroundings, but the result of their own thinking and processing. They use what they already know as a basis for making sense of any experience they encounter.

Whereas cognitive theories generally view learning as the acquisition of knowledge structures through active processing of information in terms of existing ideas that can lead to a change in the meaning of experience, constructivists argue that learning is not just the adding of new information or knowledge to the existing one but involves the restructuring of knowledge in the mental system (cognitive structure) by the learner. Knowledge is viewed as a human construction.

The constructivists believe that learners must construct meaning for themselves; that the only learning that can take place is that which is connected to the individuals' already existing knowledge, experiences or conceptualisations. What children learn is not a copy of what they observe in their surroundings but the result of their own thinking and processing. They use what they already know as a basis to make sense of any experience they encounter. The theory, therefore, focuses on what the learner has to do to construct knowledge as being the most important thing in the learning process. This has to do with the nature of the learning activities the student uses to construct knowledge. The knowledge created is built upon what the student already knows.

The theory emphasises that the student creates knowledge and that knowledge is not imposed or transmitted to the learner by direct instruction. Knowledge, then, is created with the help of the students' learning activities. What the students construct from these activities (or learning situation) depends on their:

- Motives and intentions;
- Prior knowledge or information or experience; and
- Use of prior knowledge (how they use it).

Learning is, therefore, a way of interacting with the world. As students learn their conceptions of phenomena change, and they see the world differently. The acquisition of information in itself does not bring about such change, but the way learners structure that information and think with it brings the changes. Meaning is therefore personal and is not transmitted from the teacher to the learner.

Children begin their formal study of science with ideas already in place about the natural world. Some of these ideas (children's alternative conceptions of science) are congruent with currently accepted scientific understandings, some are not. For conceptual change to occur, children must become dissatisfied (cognitive disequilibration) with their existing conception. Their existing conception, believable on the surface, must be shown to fail to explain some new observation satisfactorily. For example, the belief that heavy balls fall faster than light balls does not account for the baseball and the marble hitting the floor at the same time when dropped from the same height. The belief that the speed of a pendulum ought to depend on the weight of the bob is not supported by experimental evidence. Children have to experience for themselves occurrences that contradict their currently held beliefs before they strive to resolve the conflict between what they observe and what they believe.

There are various perspectives within constructivism based on the premise that knowledge is not part of an objective, external reality that is separate from the individual, but a human construction. Two major belief systems exist, namely, **Radical Constructivism** and **Social Constructivism**. Both perspectives describe the learner as a constructor of knowledge who actively searches for meaning, involving formation, elaboration, and testing of possible mental structures until a satisfactory structure emerges. The two are elaborated on next.

Radical Constructivism is derived from Piaget's cognitive development theory. Piaget maintains the existence of external reality to which learners must gradually accommodate their thinking. However, radical constructivism maintains that all knowledge is a human construction and that individually constructed viewpoint cannot be judged as less correct than another except on the score of their conformity with accepted norms. Its focus is not on logical thinking, but on tasks involved in school learning. According to this view, learning is an internal process that occurs in the mind of an individual and progresses through several internal reorganisations that become more encompassing and integrative. In this case, cognitive conflict occurs when one's thinking is challenged and reflections are essential learning processes. The teacher's role in this enterprise is to:

- Develop an adequate model of students' ways of viewing an idea;
- Devise situations that challenge the students' ways of thinking; and
- Help students examine the coherence in their current mode of thinking.

The implication of these roles leads to alterations of classroom configuration from teacher-directed instruction to problem-solving learning. The students should make conjectures, abstract properties, explain their reasoning, validate their assertions and discuss and question their own thinking and that of others. In this case, the teacher's role is to choose problems that can be solved in different ways and which have the capacity to probe students' thinking as they discuss problem approaches and their thinking.

Social Constructivism differs with radical constructivism in three ways: The definition of knowledge; the definition of learning; and the focus of learning (see Table 5.2). Social constructivists interpret Vygotsky's view of knowledge as constructed by cultures in a classroom. They view the classroom as a **community** charged with the task of developing knowledge (or a **community of learners**). They view knowledge as inseparable from the activities that produced it, namely, that knowledge is transactional, learning is socially constructed, and learning is distributed among the co-participants. The role of the learner is therefore perceived as an active participant in a system of practices that are dynamic.

Therefore classroom learning within the framework of social constructivism involves evolving practices in a **community of learners** including students' acquisition of cognitive tools (skills and strategies) through participation in hands-on project-based approach to instruction, collaborative learning, cooperative learning, group work, and team bonding. In the process, students may also develop big ideas which can enlarge a student's field of vision and have the potential to transform thinking. The teacher's role in the circumstance is to assist students to overcome learning obstacles by probing the limits of their understanding with difficult cases and helping them manage uncertainty. Constructivism also seeks to overcome the passivity of rote learning and to enrich the knowledge of students through experiences in thinking that reflect discipline-based practices.

The social constructivism can be applied in the transformation of schools into learning communities rather than just classrooms. This is because according to social constructivism, learning is the province of a community of participants. Every school may collectively support students in all classrooms to develop shared meaning through their participation as in attempting to acquire cognitive tools through participation in hands-on projects, or in developing big ideas in a subject like biology.

Students develop a constructivist perspective by encouraging them to work in groups, to involve in reflective dialogue, to participate in ground activities in a real-world setting, and to participate in negotiating meaning through discussion of their differing interpretations (Table 5.2).

Table 5.2: Perspectives on Constructivism

Key item	Perspective
Definition of knowledge	A product of the particular classroom or participant setting to which the learner belongs; the endpoint or product of a particular line of inquiry that is inseparable from the occasions and activities that produced it.
Definition of learning	Socially shared cognition that is a process of becoming a member of a sustained community of practice; social interaction that constructs and reconstructs contexts, knowledge and meanings.
Locus of learning	Not confined to the individuals' mind; occurs in a community of participants and is distributed among the co-participants.

Radical (or individual) constructivism and social constructivism are the two categories or perspectives of constructivism. They are slightly different in the teaching philosophy in terms of the definition of knowledge, learning, the role of learners, the role of teachers, and the role of peers as summarised in Table 5.3.

Table 5.3: Fundamental Characteristics of Social and Radical Constructivism

Key Teaching/ Learning Items	Radical/Individual Constructivism	Social Constructivism
Knowledge	- Understanding that knowledge is constructed, not transmitted - Changing the body of knowledge individually constructed in the social world - Built on what learner brings	- A product of the particular classroom or participant setting to which the learner belongs; - The endpoint or product of a particular line of inquiry that is inseparable from the occasions and activities that produced it; - Socially constructed; - Built on what participants contribute and construct together
Learning	- Active construction and restructuring prior knowledge	- Collaborative construction of socially defined

		knowledge and values
	- Occurs thro' multiple opportunities and diverse processes to connect to what is already their own.	- Occurs thro' socially constructed opportunities - Socially shared cognition that is a process of becoming a member of a sustained community of practice; - Not confined to the individuals' mind; occurs in a community of participants and is distributed among the co-participants. - Social interaction that constructs and reconstructs contexts, knowledge and meanings.
Teaching	- Challenge, guide thinking toward a more complete understanding	- Construct knowledge with students
Role of teacher	- Facilitator, guide - Listen for students' current conceptions, ideas, thinking	- Facilitator, guide - Co-participant - Construct different interpretation of knowledge - Listen to socially constructed conceptions
Role of peers	- Not necessary but can stimulate thinking, raise questions	- Ordinary part of the process of knowledge construction
Role of students	- Active construction within the mind - Active thinker, explainer, interpreter, questioner	- Active construction with others and self - Active thinker, explainer, interpreter, questioner - Active social participator

Implications of Constructivism for Teaching Biology

The learning of science in general and biology, in particular, is a process of construction and reconstruction of personal theories previously held. It is a process of continually refining existing knowledge and constructing concepts in intricately organised networks that are unique to each student. The constructivist teacher seeks to induce cognitive disequilibration by setting up situations that encourage students to question their existing beliefs and ask what is going on. This brings the existing beliefs to the surface, giving the teacher access to what is in the student's minds and thus the opportunity to help them reconstruct their beliefs in valid ways that include the new information and to make sense of them. The teacher helps to provide exploratory and predictive power-at-last, until the next observation that produces disequilibration. On this Von Glaserfeld says:

"...*if the objective is to lead the children----to some form of understanding----teachers must try to infer---what the children's concepts are and how they operate with them. Only on the basis of some hypothesis can teachers devise ways and means to orient, direct, or modify their mental operating".*

Since no two people internalise the same experience the same way, the information imparted by the teacher is not necessarily learned the same way. The teacher must learn how each student is constructing information and then help each to attach new experiences in ways that are both meaningful and convincing to them. The teacher does this by asking questions to see how students may have previously constructed information related to the topic. The teacher leads them through exploratory activities that enable them to investigate on their own and come to their conclusions as to what is happening. The teacher interacts with each student to see how each is constructing new information and helps each to formulate sound conclusions by aiding them in constructing information in ways that are both valid and meaningful to them.

The latter process could be achieved by the teacher introducing a minimal understanding of the conceptualisation as generally agreed upon by the scientific community. The child is thus compelled to relate the new phenomenon, new ideas and new experiences and observations to existing knowledge in ways that are most appropriate to the student.

Overall Implications of Learning Theories for Biology Teaching

The major implications arising from the learning theories outlined in this chapter with regard to the teaching and learning of biology are outlined below.

Design of Learning Activities

The theories indicate that '**activity**' refers to both psychomotor and cognitive aspects. **Learning activities** in biology are the tasks the learner is engaged in to gain proficiency in biological knowledge, scientific skills and attitudes planned for. Learning is fundamentally constructed as a result of the learner's activities. Learning cannot be developed without active participation by the learner. We learn by doing, and so full provision needs to be made for biological activities that facilitate learning.

Learning activities contribute to learning through the complex interactions they stimulate within the central nervous system that includes the brain. Through activity, experience and skills are acquired and new learning is better understood.

The learning activities in biology may include laboratory practical work or class experiments; discussions; field course excursions; reading and making notes; project work; listening to guest speakers; dramatization, and music among others. The learning activities should be as varied as possible to include: student/teacher controlled activities; group/individual activities; use of all the senses; and use of full range of multimedia.

The activities should encourage full participation by the learners and should be interesting and challenging enough to arouse and sustain learner's interest. The teacher selects classroom/laboratory activities such as laboratory investigation tasks, and questions to be investigated; instructional strategies to be used in teaching the content such as demonstrations, experiments and fieldwork; and resources to be used such as texts, references, films, slides, and library research. The activities should be designed keeping in mind individual differences, particularly the age of the learner, the subject matter, prior experiences of the learner, and the objectives of a given lesson. These should stimulate students' thinking and arouse their curiosity.

Creation of Motivation

Pupils can be motivated to learn biology by providing them with opportunities to explore and manipulate living things in their surroundings. This best achieved by encouraging them to solve identified problems either individually or in groups and by helping them check their success in these scientific investigations. In this way, pupils learn using intrinsic (application, exploration, manipulation and achievement) rewards. Pupils develop interest in learning biology when incentives in the form of extrinsic rewards (words of praise, encouragement, and recognition), immediate knowledge of satisfying results, cooperation, and self-competition or competition with others are used to encourage performance on various tasks. Pupils can also gain optimum interest in biology by giving them information and designing curriculum that engages them in pursuits that have everyday relevance.

Encouraging Schematic Learning

It is evident that in any new field of knowledge the schemata first formulated have a lasting consequence on future learning in the field (Child, 2004).

The first task for any biology teacher is to discover and define the fundamental schemata required to enable the most productive assimilation and then proceed to form a familiar to unfamiliar knowledge. The teacher may do this by encouraging pupils to organise past actions which become reference points for interpretation and development in future learning. In this way, the teacher is expected to encourage pupils to construct their own meanings of biological concepts, develop their own understandings of the uses and nature of the processes, and apply biological knowledge to their lives in ways that are meaningful to them. Schematic learning is more efficient than rote learning as the learner builds systematically on previously acquired learning. Rote learning in the absence of understanding precludes the logical acquisition of further meaningful knowledge. However, rote learning may be used to learn any symbolic form new to the learner, such as chemical symbols, basic mathematical or physical science equations, and structure and function, taxonomic tasks, and chemicals of life in biology that require a combination of rote learning and logical build-up from previous knowledge. According to Child (2004), rote learning seems to have its own brain bases.

The biology teacher is expected to understand the intellectual limitations and advantages available to pupils as they progress through the stages of cognitive development. Combining intellectual capability with individual construction will ensure a meaningful biology programme, one that will meet the goals and objectives of science education.

Use of Reinforcement

Biology teachers should insist on realistic goals or targets for the pupils thus ensuring some measure of success for each pupil. This increases the likelihood of reinforcement. In this case, it will be easier for pupils to be reassured about levels of success. For it to be a really effective reinforcer in achievement, knowledge of results must follow quickly upon completion of a task in biology. Then it will have maximum influence on future performance.

Encouraging Metacognition

The ability to learn how to solve problems of a given kind can be developed with sufficient practice on tasks of similar nature. This ability is known as **learning set,** or **learning how to learn** or **metacognition.** These are habits of doing things: of thinking and characteristic ways of tackling problems. Learning how to learn, a subject as well as acquiring the rules to be applied to the subject matter, is a crucial learning activity. Some individuals function more effectively than others as independent learners due to differences in their metacognition (knowledge of how they think). The key components of metacognition (knowledge of how one thinks, including knowledge of strategy components and their functions) include:

- **Analysis** - identifying factors relevant to achieving a goal;
- **Planning** - formulating an initial statement of intended action;
- **Implementation** - use of memory and comprehension tactics in a skillful manner;
- **Monitoring** - periodically determining the adequacy of strategy implementation; and
- **Modification** - making changes in the analysis, planning, and implementation if needed.

Students should be encouraged to understand the interrelationships of concepts and sub-concepts by developing concept maps as a major learning strategy.

Learning Styles

It is evident that students learn more and retain more when they learn in a manner that is comfortable to them. People learn and process information in different ways, namely, visual, auditory, kinesthetic learning styles. Visual learners learn best by seeing; auditory learners learn best by hearing, while kinesthetic learners learn best by touching and feeling. For each learner, one of these styles is stronger than the other two, although these are also functional. The teacher needs to plan activities appropriate for each learning style so that all children can pursue their investigations in styles that are comfortable, challenging and meaningful to them. The process-oriented inquiry is the most appropriate vehicle for the accommodation of the three learning styles because pupils are able to pursue their explorations in the learning styles that are most comfortable to them (Martin, 2000).

While it is important to consider learning styles in the teaching of biology, it is also necessary to realise that the 'locus of control' phenomenon described by Martin (2000) could greatly contribute to the learning outcomes. This trait concerned with whether people attribute responsibility for their own success or failure in academic performance to internal or external factors. Some have a predominantly internal locus of control while others have mainly external locus of control. Those with the former believe their success or failure depends on their own abilities and efforts such as their behaviour, persistence, inquisitiveness and intelligence (internal locus of control) while the latter believe that their success or failure depends on external factors such as luck, other people's action, or the difficulty of the situation (external locus of control).

Students with internal locus of control rely on both sensing/feeling or on intuition/thinking. They perceive by sensing the need to see, hear, feel, taste, or smell a stimulus for it to be registered in the brain or by thinking/intuition, employing rules of logic as 'if---then' clauses and syllogistic reasoning. The pupils thus demonstrate a higher rate of academic achievement than their counterparts. On the other hand, pupils with an external locus of control believe they have little or no control of their performance or achievement and have no motivation to engage in exploratory work or to pursue investigations on their own. They thus demonstrate a low rate of academic achievement. Teachers are expected to design activities that can encourage such pupils to manipulate and be engaged in a dialogue intended to help them see their roles (Martin, 2000). The process-oriented inquiry can enable pupils to develop their own investigations to answer questions they raise from the observations they make. In this way, pupils take charge of their own learning.

Cognitive Demand

Cognitive learning theories indicate that concept learning should be the focus of attention as it is central to the development of an understanding of the biological world. The most important aspect of the teacher's planning should, therefore, be the selection and ordering of concepts to be learned in biology. Some concepts are more abstract and sophisticated as the case of secondary, correctional and theoretical concepts than others such as classificational and primary concepts e.g. insect.

To give meaning to concepts in biology, it is desirable to start with primary concepts through laboratory and fieldwork to enable pupils to have experiences with real objects needed to give meaning to such concepts as a cell, mitochondrion, photosynthesis, buttercup, food chain, and climax vegetation. Once these primary concepts are established to some degree of cognitive differentiation, they can be used in new combinations to form secondary concepts such as ecosystem, evolution and metabolism. In this case, inquiry learning should be used to provide pupils with opportunities designed to test for concept meanings from which other desired behaviours derive such as processes and attitudes. Hands-on experiences for pupils are useful wherever new primary concepts are to be learned. However, older pupils may require substantially less concrete experience since their past experience permits them to discern essential attributes of events or objects that are necessary for primary concept development.

Planning

Biology teachers are expected to give a clear framework on which planning for instruction can be based. They should plan instruction systematically by constructing objectives clearly and defining skills accurately. This permits prerequisites to help in the learning of terminal skills and in selecting suitable concepts for the purpose. In this way, it is possible to structure content or knowledge to suit the learners.

Conclusion

This chapter has provided vital information on the theories that inform the learning of biology. Learning is seen as the process by which learners acquire and retain knowledge, attitudes, skills and capabilities that cannot be attributed to inherited behaviour patterns or physical growth. It is evident that the learning of biology is a process of the learners' construction and reconstruction of personal beliefs and conceptions previously held. It is a process of continually refining existing knowledge and constructing concepts in intricately organised networks that are unique to each learner. .Whereas the current practices lean towards the constructivist theory of learning, information from other theories is still important because some teaching and learning practices are predicated on them. Research on the learning process continues and the results are having some effect on the approaches to the teaching of biology.

References

Biehler, Robert F and Snowman, Jack (1990) *Psychology Applied to Teaching: Study Guide* 6th Edn Boston: Houghton Mifflin

Child, Dennis (2004) *Psychology and the Teacher*. 7th Edn. London: Continuum

Kauchak, Don and Eggen, Paul (2011) *Introduction to Teaching: Becoming a Professional* 4th Edn. New Jersey: Pearson Educational Inc.

Martin, David J. (2000) *Elementary Science Methods* 2nd Edn. Belmont: Wardsworth/Thomson Learning

Schultz, Jerome (1991) *Instructor's Resource Manual: Educational Psychology* 2nd Edn. Boston: Houghton Mifflin

UNESCO (1977) *New Trends in Biology Teaching*, Volume IV. Paris: UNESCO

Woolfolk, Anita (2001) *Educational Psychology*. New Jersey: Pearson Educational Inc.

THE EFFECTIVENESS of teaching and learning is, therefore, an interactive process involving students and teachers

Photo: UN WOMEN/FLICKR

6

CHAPTER SIX

LEARNING IN BIOLOGY

◆ Approaches to Learning in Biology ◆ Surface Learning ◆ Deep Learning ◆ Teaching and Learning in Biology ◆ Effective Teaching ◆ Concept Learning in Biology

Introduction

The learning of science is a process of the learners' construction and reconstruction of personal beliefs and conceptions previously held. It is a process of continually refining existing knowledge and constructing concepts in intricately organised networks that are unique to each learner. It is the process by which learners acquire and retain knowledge, attitudes, skills and capabilities that cannot be attributed to inherited behaviour patterns or physical growth (Farrant, 2002). This chapter outlines some of the conditions and situations that facilitate the learning of biology.

Approaches to Learning in Biology

Learning is constructed as a result of the learner's activities. Learning activities contribute to two broad approaches to learning: **Surface Learning** and **Deep Learning**. Deep and surface approaches to learning describe the way students relate to a teaching and learning environment. Good teaching leads to deep learning, while poor teaching promotes surface learning in which students use low-order thinking skills.

Surface Learning

Surface learning is the type of learning caused by generally inappropriate learning activities that constitute students' low cognitive engagement. The activities do not motivate the learner to use prior knowledge to construct meaning and thus create new knowledge. The learner does not construct adequate meaning from a learning situation for lack of adequate understanding. The activities yield fragmented outcomes.

Surface learning occurs where the learner is introduced to low-cognitive activities instead of the more required higher-cognitive activities. This situation arises because the teacher is in a hurry to complete the task with minimum effort while appearing to meet the course requirements. Examples of surface learning activities include:

- Learning selected content through memorisation instead of understanding it fully;
- Describing events, situations and processes without providing explanations;
- Listing points without providing explanations;

- Copying notes from textbook or chalkboard verbatim without reflection;
- Quoting secondary references as if they were primary ones; and
- Quoting back from the notes given by the teacher instead of providing informed explanations.

The factors that promote students' adoption of surface learning include:

- An intention only to achieve a minimal pass;
- Insufficient learning time or too high a workload;
- Lack of opportunity to engage as active participants in the learning process;
- Misunderstanding cognitive requirements, such as thinking that factual recall is adequate;
- Lack of sound knowledge foundation;
- Inability to engage fruitfully in high cognitive activities due to a lack of basic cognitive skills;
- High anxiety; and
- Inability to understand particular content at a deep level.

The factors that encourage teachers to adopt surface learning approach in their teaching are:

- Teaching piecemeal, not bringing out the holistic structure of the subject matter;
- Interested only in separate facts without relating them to the whole;
- High workload;
- Insufficient preparation of teaching material;
- Providing insufficient time to fully engage students in the tasks;
- Use of unsuitable instructional strategies and materials;
- Emphasising coverage of content at the expense of depth; and
- Having low expectations of success for the learners.

The first step in improving teaching is to avoid those factors that encourage a surface approach.

Deep Learning

Deep learning is the learning involving the use of appropriate learning activities that engage students in high-level cognition. The learners try to focus on the underlying meaning as well as the main ideas, themes, principles and application in order to understand the big picture of the learning task. This requires a sound foundation of relevant prior knowledge on the part of the students to help them learn the details as well as understanding the big picture. The big picture is not captured without identifying the associated minor details.

When the deep approach is used, students have positive feelings about what they learn such as raising questions about what they learn, developing interest in the task and a sense of self-esteem.

Examples of deep learning activities include theorising; applying knowledge to new situations; explaining; raising questions; hypothesising; and giving alternative ideas/explanations.

The factors that encourage students to adopt deep learning are:

- Willingness and ability to engage in the task fruitfully;
- The determination to perform well in the given task;
- Adequate and relevant prior knowledge;
- The ability to engage the learning task at a high cognitive level, which requires a well-structured knowledge base;
- High level of success expectation for the students; and
- Ability to abstract and work with the big picture rather than with unrelated detail.

The factors that encourage the teacher to adopt deep learning approach in their teaching include:

- Ability to explicitly bring out the structure of the subject matter;
- Using instructional strategies that elicit an active response from students, e.g. by questioning, presenting problems, rather than teaching to expound information;
- Ensuring that learners are actively engaged in the learning process and using high thinking skills;
- Teaching by building on what students already know;
- Identifying and removing students' misconceptions;
- Assessing for structure rather than for independent facts;

- Encouraging a positive working atmosphere for the students so that they can make mistakes and learn from them; and
- Emphasising depth of learning, rather than breadth of coverage.

Teaching and Learning in Biology

Teaching and learning are closely intertwined. Learning involves internal processes in the learner; but the process is often influenced by external events or stimuli from the environment, usually in form of direct verbal communications from a teacher, or indirect communication from a textbook, apparatus or some other source. The external events, when they are planned for the purpose of supporting learning, are called **instruction**. As a **manager of instruction**, it is the teacher's responsibility to plan, design, select and supervise the arrangement of these external events, with the aim of activating the necessary learning process (Biggs, 2003).

Several factors are involved in the teaching and learning process. Biggs (2003) posits the 3P model of teaching and learning, which describes three points in time at which learning and teaching interact to cause learning by the students. The three are presage, process, and product factors.

The **Presage** factors include the nature of the student in terms of the relevant prior knowledge the student has about the topic, interest in the topic, student ability, commitment to school, and motivation. They also include the teaching context in terms of what is intended to be taught (objectives), how it will be taught (instructional strategies) and assessed, the expertise of the teacher, and the climate or ethos of the classroom and the school itself.

The **Process** denotes the type of engagement or learning activities by the student, whether deep or surface learning. This depends on the manner of interaction of the presage factors. A student with little prior knowledge of the topic is highly unlikely to use a deep learning approach (i.e. high-level cognition), even where the teaching is expert. Similarly, a student who already knows a great deal of what is to be taught and is very interested in the topic is likely to engage in a high-level cognition or deep learning.

The **Product** refers to the learning outcome including different levels of knowledge, skills and attitudes and is determined by many interacting factors. The student and teaching related factors jointly determine the learning type (surface or deep) the student adopts for a given task, which in turn determines the quality of the outcome.

The effectiveness of teaching and learning is, therefore, an interactive process involving students (with their abilities, motives, personality) and teachers (with their abilities, motives and personality). This means that what works for one class may not necessarily work for the other. Collectively, these background factors determine the cognitive processes the students are likely to use, which in turn determines the detail and structure inherent in the learning outcomes.

Effective Teaching: Constructive Alignment and creation of a Learning Community

Good teaching is getting most students to use **the higher cognitive** level processes that the more academically oriented students are likely to use spontaneously. The higher cognitive level processes include theorising, applying, generating new ideas, reflecting, relating, explaining, and problem-solving. The **low-level cognitive** processes include memorising facts, note-taking from teachers or directly from textbooks, and describing activities and events in order to learn. The low-level cognitive processes are encouraged through a surface approach, where passive teaching strategies are employed, while the higher cognitive level processes are encouraged through a deep approach where active teaching strategies are employed.

According to Biggs (2003) the critical components of teaching and learning are as follows:

- The curriculum that we teach;
- The teaching methods that we use;
- The assessment procedures that we employ and methods of reporting results;
- The climate that we create in our interactions with the students; and
- The institutional ethos including the climate, the rules and procedures we have to follow.

These components interact as a system. Teaching is aligned with each of these components in the system to promote deep learning. Imbalance in the system will lead to poor teaching and surface learning. Non-alignment is signified by inconsistencies, unmet expectations, and practices that contradict what is believed in.

For example, a teacher of biology works in a school with its policies, rules regulations, aspirations and culture collectively referred to as its ethos. The teacher has no control over these but has to try as much as possible to operate within this framework. When the teacher goes to teach, the teacher sets up and controls the classroom climate. The kind of atmosphere the teacher creates-(whether authoritarian, friendly, cold, or warm) can markedly influence the effectiveness of a teaching approach in biology. The teacher has also to align the curriculum, the teaching techniques and strategies, and the assessment procedures for effective teaching. Biggs (2003) points out that when there is alignment between what we want, how we teach and how we assess, the teaching is likely to be much more effective than when it is not.

In a classroom situation, there are usually two groups of students in terms of learning ability, namely, students who are academic in orientation and non-academic students. The academically oriented students spontaneously carry out the higher levels of learning activities more or less independently of the teaching. For them, lecturing works perfectly well. The non-academic students (usually the majority) need more support in order to carry out these higher-level activities since they rely on low-level cognitive activities in order to learn. Teaching is expected to provide the support they need. To narrow the gap between the two types of students, teachers must use more interactive teaching strategies such as problem-based learning, collaborative learning, group discussion, simulations, among others that assist the non-academic ones to engage in higher cognitive activities that the 'better' academic ones do under non- interactive methods such as lecturing.

The act of teaching that narrows the gap between their levels of active engagement in learning is called **constructive alignment.** The constructive alignment model aligns curriculum objectives, teaching/learning activities (TLAs), and assessment tasks (see Biggs, 2003). The first aspect of the model is what the teacher constructs and the second aspect is how the student reacts. This model is designed for teaching calculated to encourage deep engagement. Through **constructive alignment,** more students actively participate, that is, co-operate and share in the learning process. They will learn and remember because they are focusing attention on a task. **Constructive alignment** is thus a combination of constructivist theory and aligned instruction. This model is designed for teaching with a view to encouraging deep engagement. This is designed by considering:

- The specific level of understanding of the content in question by stipulating the appropriate action verbs for the objectives;
- Specific target activities to be performed by the learners as per the verbs stipulated;
- Specific teaching techniques and strategies to be used that engage pupils in deep learning; and
- Specific assessment tasks to address, in order to judge if or to what extent, the students have been successful in line with the instructional objectives.

Constructive alignment is actually a combination of the constructivist understanding of the nature of learning and the aligned design for teaching. In aligned teaching, there is maximum consistency throughout the system. The curriculum is stated in terms of clear objectives, which states the level of understanding required rather than simply a list of top-ics to be covered. The teaching approaches are selected that are likely to realise the objects. The teacher gets students to do the things the objectives indicate. The assessment tasks address the objectives so that the teacher tests to see if the students have learned what the objectives state they should be learning. The hope of the system is that students will engage in appropriate learning activities (deep learning) and thus construct their knowledge.

The objectives are stated in terms of action verbs, which indicate the constructive activities that are most likely to achieve the desired outcomes for the biology topic or unit in question. For example, verbs denoting high-level cognitive engagement include: theorise, reflect, generate, apply, while low-level engagement is denoted by memorising, describe, and recognise, among others.

The biology content is then operationalised in terms of the levels of understanding in a hierarchy that corresponds to the grading system used. The categories used to define the quality of learning and understanding. The TLAs are then designed that are likely to encourage students to engage in higher-order cognitive abilities. Finally, assessment tasks that inform the teacher whether and how well each student can meet the criteria expressed in the objectives. The objectives, teaching and assessment are now aligned and most students will be expected to engage with appropriate activities or in deep learning.

Constructive alignment, therefore, engages the students in active learning and the teacher simply acts as a mediator between the student and a learning environment that supports the appropriate learning activities. Constructive alignment is possible where the learning situation engages the learners to use all their senses; to share and collaborate; to listen and respond; to reflect and articulate, and to connect and create. This is the sure way of creating a **classroom learning community.** The more the student is focused on the learning task, the higher the percentage of learning and remembering. It is a well-known fact that we learn and remember:

- 10% of what we **hear;**
- 15% of what we **see;**
- 20% of what we both **see** and **hear;**
- 40% of what we **discuss** with others;
- 80% of what we **experience** directly or **practice;** and

- 90% of what we attempt to **teach** others.

Barriers to Constructive Alignment

The teaching of biology by many teachers is not aligned. There are several reasons for this:

- Traditional transmission theories of teaching ignore alignment. For example, a common method of determining student's grades depends on how students compare with each other (norm-referenced), rather than on whether an individual's learning meets the objectives (criterion-referenced). There is no inherent relation between what is taught and what is tested;
- Some administrative factors, such resource limitations, which dictate large classes with mass lecturing and multiple-choice testing, make alignment difficult; and
- Many teachers are not aware of the significance of aligned instruction.

Concept Learning in Biology

Some concepts have fewer attributes and these are easier for pupils to learn. Others have for too many interrelated attributes and are difficult for them to learn. It is important for the biology teacher to identify the categories of concepts so that proper planning is made to teach them.

Types of Concepts

a) Taxonomic or Classification Concepts

These concepts are identified by virtue of possession of a common set of properties. No value or quantity is attached to them.

Examples:

Square, mammal, Red, transitional element, Fish, insect, Two, metal, Mole, and mammal. These concepts are not learned in the same way. Some are learned by direct observation, e.g. square, red; others require ground knowledge to be learned e.g. mammal – requires knowledge of the animal, far, milk, giving birth, suckling. The transitional element needs abstraction including other concepts.

b) Quantitative Concepts (Non-Taxonomic)

These concepts involve quantified variables – i.e. they have a numerical value.

Examples:

The concepts of length, volume, momentum, mass, acceleration; all have value. For instance, volume occupies space, 3=dimensional, LxWxH.

c) Correlational or Relational Concepts

These concepts are multi-dimensional and require the linking of facts together to describe them.

Examples:

Energy, Force, density, motion, velocity, mole concepts.

- Force is seen as a product of mass and acceleration (F=Ma);
- Density is seen as a quotient of mass and volume;
- Energy embraces a variety of meanings: Kinetic, gravitational, potential, electric, heat, radiation, mass-energy etc., which are related together through a highly abstract conservation law; and
- The concept of energy is fully understood or defined if all these various supportive concepts and law statements which relate various concepts are indicated. For children to understand energy the accompanying conservation law is necessary.

d) Theoretical Concepts

These are abstract concepts since they are postulates; they are not capable of being observed or experienced directly in themselves. Students find such concepts difficult to learn without the teacher's prompt intervention.

Examples: atoms, molecules, electrons.

Concept Learning

The more concrete concepts are usually learned through verbal instruction. This helps the student to attach an identity to an object or idea. For example, in learning about a mammal, the students are shown the actual mammal known to them or the photographs of such a mammal. They then identify the common characteristics of mammals. They even internalise this concept faster if other animals that are not mammals (non-examples) are shown to them. The features that characterise a mammal as a concept are:

- The hair on the body;
- Mammary glands;
- Four limbs;
- Suckling the young; and
- Giving birth to young

The more abstract concepts are more difficult to learn. Such concepts have the following features:

- Theoretical in nature - concepts the teacher may not easily arrange for concrete experiences;
- Have a wide range of inter-relationships (large concept maps); and
- Require a lot of mathematical applications-e.g. surface area, surface area to volume ratio.

The teacher is expected to help the student to "visualise" or "concretise" or "conceptualise" such concepts for ease of internalisation. This may be achieved through the use of visual aids and experiments or investigations. Experiments or investigations provide the foundational experiences that lead to reflective observation (observing and explaining what happens). This is when the students can conceptualise or understand the abstract concept.

The teacher may also use concept maps, flow charts, classification charts, and concept trees or webs to indicate the relationships between the more abstract concepts. Concept maps give better detail as they show the interconnectedness of concepts at similar hierarchical levels, whereas the others are linear, progressing from one point to another without showing the linkage between branches. For example, the roots of the concept of respiration may be determined by breaking it down in a concept map to indicate the various concepts and sub-concepts that must be learned by the student to make it more meaningful and to aid remembrance.

In teaching for concept formation, the teacher is generally expected to:

- Gather the facts relevant to the concept;
- Link the facts to make some order or pattern;
- Find out some similarities and differences among the facts; and
- Identify concepts and non-concepts.

The information assists the teacher to prepare adequately to teach for concept formation. Establishing a concept usually involves:

- Learning and understanding through the use of more exemplars and non-exemplars;
- Generalisation – extending the bounds of the concepts so that examples other than those which have already been met can be recognised;
- Discrimination – excluding irrelevant attributes and labelling (describing a concept by word);
- Providing a wide range of experiences both concrete and abstract;
- Giving an accurate verbal description;
- Teach from concrete to abstract concepts and from known to the unknown;
- Use of visual aids; and
- Identification of interconnectedness of concepts.

Box 6.1

Discussion Questions

1. Study the concept map in Fig 8.2. Add the link words between each box to describe the concepts that relate to that key concept of respiration; and
2. Provide the possible advantages of using concept mapping in the teaching and learning of abstract concepts.

Conclusion

This chapter has emphasised that learning is constructed as a result of the learner's activities, which either contribute to **Surface Learning** or to **Deep Learning**. Deep and Surface approaches to learning describe the way students relate to a teaching and learning environment. Good teaching leads to deep learning, while poor teaching promotes surface learning in which students use low-order thinking skills. Teachers are encouraged to enhance deep learning as opposed to surface learning. This is achieved if biology teachers adopt constructive alignment strategies that balance the abilities of learners in a given class. The teacher adopts teaching strategies that favour the less academically oriented students while safeguarding the academically oriented ones. An important element in this strategy is the use of concept learning particularly concept mapping.

References

Biggs, John (2003) *Teaching for Quality Learning at University* (2nd Edn). Berkshire: Open University Press

Gagne, R.M and Driscoll, M.P (1988) *Essentials of Learning for Instruction* (2nd Edn) New Jersey: Prentice Hall.

Marton, F (1981) 'Phenomenography – describing conceptions of the world around us.' Instructional science, 10: 177-200

Marton, F and Saljo, R (1976) On qualitative differences in learning: I: outcome and process, *British International Journal of Educational Research Journal of Educational Psychology*, 46: 4-11

Marton, F, Dall'Alba, G and Beaty, E (1993) Conceptions of learning, 19: 277-300

Twoli, N. W (2006) *Teaching Secondary School Chemistry*. Nairobi: Nehema Publishers

EFFECTIVE CONSTRUCTIVE teaching must be anchored in an instructional philosophy

Photo: STAR FOUNDATION/KRISTIAN BUUS/FLICKR

7

CHAPTER SEVEN

INSTRUCTIONAL PHILOSOPHY IN BIOLOGY

◆ Instructional Philosophy ◆ Effective Constructive Teaching of Biology for Meaningful Learning ◆ A Personal Instructional Philosophy in Biology

Introduction

Teaching is guided by theories and philosophies of education. Philosophy is a search for wisdom and seeks to provide an insight into the meaning of knowledge and how this knowledge is acquired. Every teacher is guided by a set of related beliefs and ideas that influences what and how students are taught. We refer to this as the philosophy of education. Philosophies of education are based on the way the schools and teachers resolve the various philosophical questions that have puzzled thinkers since the time of the ancient Greeks. The key questions in education are:

- What is the purpose of schooling (Aims of education)?
- What should schools be teaching (The curriculum)?
- How should teachers teach (Instruction or teaching)?

Philosophy and theory are different but related. A theory is a set of related principles that are based on observation and are used to explain events, behaviour and phenomena we observe in our day-to-day lives in the world around us. Philosophies, on the hand, suggest the way events and behaviours ought to be. The function of philosophy is to help teachers understand how past experts have thought about teaching and provides a framework for thinking and professional practice.

This chapter intends to help the reader delve into the issue of a philosophy of instruction in biology as a discipline of knowledge.

Instructional Philosophy

Philosophy of education is a set of ideas and beliefs that guides teachers to understand more about teaching and teachers' actions in the classroom and provides a framework for thinking about educational issues. Instructional philosophy is a set of ideas and beliefs about what teachers should teach, for what purpose they should teach and how best the teachers should teach. The philosophy of education that guides the current best practices of teaching and learning in the 21st century is engrained in the constructivist theory of learning.

The basic idea of the constructivist theory is that learning is an active, constructive process; the learner is viewed as information constructor. This is a view of learning that asserts that to make sense of their experiences, students construct their own understanding of the content they study instead of having that understanding transmitted to them by someone else. This has philosophical (epistemological) connections: the way people come to know what they know is to construct the knowledge for themselves. Learners are thinking humans that are mentally active. Constructivism is, therefore, the view that emphasises the active role of the learner in building understanding and making sense of information. New information is linked to prior knowledge; these mental representations are subjective.

The constructivists, therefore, believe that each learner must construct meaning by themselves based on the already existing knowledge, experiences or conceptualisations. What children learn is not a copy of what they observe in their surroundings but the result of their own thinking and processing. They use what they already know as a basis to make sense of any experience they encounter. The theory, therefore, focuses on what the learner has to do to construct knowledge. This has to do with the nature of the learning activities the student uses to construct knowledge. The theory emphasizes that the student creates knowledge; knowledge is not imposed or transmitted to the learner by direct instruction. Knowledge is thus created by the students from learning activities.

Constructivists argue that learning is not just the adding of new information or knowledge to the existing one but involves the restructuring of knowledge in the mental system (cognitive structure) by the learner. Knowledge is viewed as a human construction that is individually constructed. Learning is, therefore, a way of interacting with the world. As we learn, our conceptions of phenomena change, and we see the world differently. The acquisition of information in itself does not bring about such change, but the way we structure that information and think with it does. Meaning is therefore personal and is not transmitted from the teacher to the learner.

The teacher's role in this discourse is to:

- Develop an adequate model of students' ways of viewing an idea;
- Devise situations that challenge the students' ways of thinking; and
- Help students examine the coherence in their current mode of thinking.

The implication of these roles leads to alterations of classroom configuration from teacher-directed instruction to problem-solving learning. The students should make conjectures, abstract properties, explain their reasoning, validate their assertions and discuss and question their own thinking and that of others. In this case, the teacher's role is to choose problems that can be solved in different ways and which have the capacity to probe students' thinking as they discuss problem approaches and their thinking. A student is involved in active construction with others and self and is an active thinker, explainer, interpreter, questioner and social participator.

The Key Features of Constructivism

The conceptions and beliefs associated with constructivism include the following:

Conception of Knowledge

- Knowledge is a product of the particular classroom or participant setting to which the learner belongs;
- Knowledge is the endpoint or product of a particular line of inquiry that is inseparable from the occasions;
- Knowledge is constructed, not transmitted and activities that produced it;
- Knowledge is socially constructed, but built on what participants contribute, construct together; and
- Knowledge is a socially shared cognition that is a process of becoming a member of a sustained **community of practice**.

Learning Process

- Students engaged in active construction and restructuring of prior knowledge;
- Occurs thro' multiple opportunities and diverse processes to connect to what is already their own;
- Learning is a collaborative construction of socially defined knowledge and values;
- Learning occurs thro' socially constructed opportunities;
- Learning is a social interaction that constructs and reconstructs contexts, knowledge and meanings;
- Learning is part of social negotiation and shared responsibility; and
- Learning is not confined to the individuals' mind; it occurs in a **community of participants** and is distributed among the co-participants.

Teaching

- Teaching is student-centred instruction, including problem-solving, collaboration;

- Teaching involves the use of complex, challenging learning environments and authentic tasks;
- Students work in groups; ground in activities in a real-world setting;
- Students negotiate meaning through discussion of their differing interpretations;
- Teaching involves the use of skill in establishing thoughtful and reflective dialogue;
- Multiple representations of content; and
- Teaching involves guiding and challenging thinking toward a more complete understanding.

The Teacher

- Teacher as facilitator, guide or co-participant will construct a different interpretation of knowledge;
- The teacher listens to socially constructed conceptions;
- The teacher listens for students' current conceptions, ideas, thinking; and
- The teacher acts as a guide, facilitator.

Student

- The student is involved in active construction with others and self;
- The student is an active thinker, explainer, interpreter, and questioner; and
- The student is an active social participator.

Effective Constructivist Teaching of Biology

Effective constructive teaching must be anchored in an instructional philosophy. This is referred to as constructive instructional philosophy. The philosophy promotes effective learning of biology by the students. The set up involves helping students to make sense of concepts and ideas in biology they learn. Students are encouraged to put those ideas together in their own minds and in their own ways. This is the process of personal knowledge construction. Instruction that promotes knowledge construction by students is often described as constructivist. Among the key features of instruction that contribute to a constructivist orientation are indicated diagrammatically in Figure 7.1.

Goals and Objectives: The constructivist teacher is expected to customise the broad goals of teaching biology such that the instructional objectives or expected learning outcomes focus more on increasing students' deep learning through active engagement. This approach promotes understanding, retention and ability to apply biological ideas in school and their everyday lives and interests in biology.

Biology curriculum: While the biology curriculum is outlined in the syllabus, the constructivist teacher is expected to align the curriculum to the spiral curriculum mode that builds systematically over time, from simpler to more complex ideas, and from shallower to deeper understanding. This should be done through schemes of work irrespective of the organisation adopted in the curricula materials that may ignore the benefits of spiralling the curriculum.

The teacher is also expected to use hands-on and minds-on (reflection, analysis, interpretation, etc.) experiences to help students develop meaningful experiences. Relevant instructional materials augment the curriculum and help students develop evidence-based reasoning to support or refute ideas in an effort to develop knowledge and scientific skills.

Biological Phenomena: The constructivist teacher is expected to identify biological phenomena, concepts and ideas from the biology curriculum to form the basis on which to engage students in raising questions and seeking for solutions to the questions before introducing them to generalisations and theories. Teaching should also provide rich sources of observational data with hands-on activities when appropriate, using films, pictures, websites, videos, models, and simulations, among others. This action helps in laying bare the biological phenomena in question.

Constructive alignment: Students in a biology class are in two categories: Academically inclined and non-academically inclined ones. The former learn faster even with non-interactive teaching approaches like a lecture, while the latter learn better with the use of more interactive instructional approaches. To narrow the gap between the two types of students, teachers must use more interactive teaching strategies such as problem-based learning, collaborative learning, group discussion, and simulations that assist the non-academic ones to engage in higher cognitive activities that the 'better' academic ones do under non- interactive techniques such as lecturing. The act of teaching that narrows the gap between students' levels of active engagement in learning is called **constructive alignment.**

In **constructive alignment,** the teacher aligns curriculum objectives, teaching/learning activities (TLAs), and assessment tasks so as to encourage deep learning or engagement by which more students actively participate, that is, co-operate and share in the learning process (see Biggs, 2003, pp24-31). Constructive alignment is the instruction designed to encourage deep learning or engagement. **Constructive alignment** is thus a combination of constructivist theory and aligned instruction.

In aligned instruction, there is maximum consistency throughout the system. The curriculum is stated in terms of clear objectives, which state the level of understanding required rather than simply a list of topics to be covered. The teaching approaches are selected that are likely to realise the objectives. The teacher gets students to do the things the objectives indicate. The assessment tasks address the objectives so that the teacher tests to see if the students have learned what the objectives state they should be learning. The hope of the system is that students will engage students in appropriate learning activities (deep learning) and thus construct their knowledge.

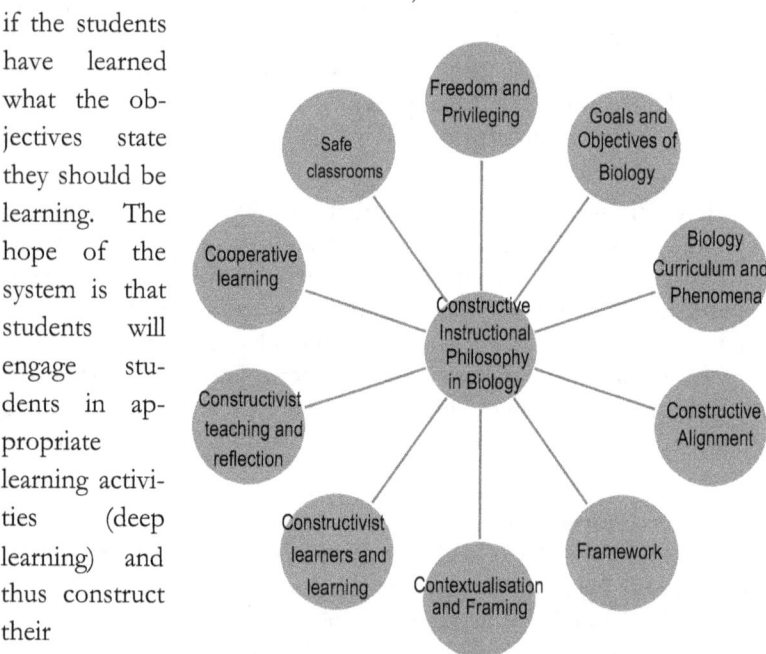

Figure 7.1: Components of Effective Constructivist Teaching

Contextualisation of Lesson content: The constructivist teacher is expected to contextualise each lesson. The biological phenomenon or content for each lesson should provide the context in which the ideas are investigated and developed by the students. It also provides the scientific skills, the motivation and the questions to be raised and answered.

Framework: The constructivist teacher is expected to design a framework in which meaningful teaching is organised. The schemes of work prepared from the biology syllabus provide a framework in which lessons are organised into a flowing and coherent entity that aims at building bigger ideas over time. Sequence and connection are important in the construction of knowledge over the learning period.

Lessons: Constructivist teaching involves designing each lesson to illustrate an important biological idea or phenomenon and implemented in a way as to give an outcome that challenges the students' assumptions and beliefs. This involves guiding and challenging thinking toward more complete understanding; it means engaging students in deep learning.

Framing: The constructivist teacher is expected to prepare a systematic blueprint for teaching intended to help the learners to construct their new biological knowledge. This is referred to as framing. The constructivist teacher is therefore expected to sequence each lesson (from set induction through to closure) to include what students learned in the previous lesson and proceeds through a series of connected experiences that build a deeper understanding of an important idea or phenomenon over time.

Constructivist Teaching: Before the lesson content is taught, the constructive biology teacher is expected to find out what the students already know about the topic through whole class discussion. The teacher elicits ideas from the students so as to identify any gaps and alternative conceptions that would be useful in the teaching and learning process in the lesson. It is equally valuable for the teacher to identify among students' prior knowledge persistent alternative conceptions (misconceptions) that impede learning and need to be specifically challenged in the context of the lesson.

Alternative conceptions are usually persistent and difficult to erase through conventional teaching techniques. The teacher is expected to identify such misconceptions and attempt to use various strategies to challenge alternative conceptions. Anchoring concepts, bridging strategies and analogies can help learners move from their alternative conceptions to more scientific conceptions.

Learners and Constructivist Learning: Learning for understanding is an active process (active mentally, minds-on) and is time-consuming and **effortful**. Learning and understanding are strongly influenced by **prior knowledge and experiences**. There are limitations to working memory for it can typically handle less than seven ideas. The use of learning resources such as pictures (visual, graphics) and/or stories help learners remember events, ideas and relations between observations and derived inferences. This means that multiple forms of representation help provide more access to ideas.

In each lesson, the constructivist teacher is expected to recognize that the learners must **construct** their own personal understanding of each topic or phenomenon. This happens if the teacher helps learners to engage in sense-making activities by first considering concrete, real, familiar experiences before abstract ones. Learners readily construct their own ideas, some of which are scientifically correct and some of which are alternative conceptions that are contrary to scientific thinking. The teacher makes an effort to engage them in activities that challenge what they already know. In this way, they use and apply their knowledge and skills in a variety of contexts so as to develop evidence-based reasoning to support or refute their scientific ideas.

To engage students actively in the lesson, the students conduct **investigations.** This involves identification of the problem, raising questions, the design of the investigation, collection of data, recording of data, interpretation of the data, linking data to everyday knowledge, and conclusion. Thus, the important questions to be raised that encourage students to conduct investigations and to remain focused on the lesson need to be identified for the lesson. The students should be asked to predict or make hypotheses regarding the outcomes of the investigations they will perform. This allows them to construct a mental model and to propose the outcome. This encourages them to look out for the outcomes during the actual lesson.

Constructivist learning is promoted when:

- Ideas are contextualised;
- Learners engage in relevant, interesting, concrete experiences which are easily defined;
- Questions emerge from students' observations that are not adequately answered by their prior knowledge;
- Learners freely discuss, share and compare their ideas;
- Sufficient time and support are provided for organising and constructing knowledge in systematic ways;
- There are opportunities for review of one's ideas by self and others, and for one's revision of those ideas;
- Successive ideas are linked together in a larger framework; and
- Appropriate tools that maximise learning are provided to the students.

Cooperative learning: Cooperative learning begins on a humble note as group work and gradually develops to create a community of learning in which groups share ideas and work towards a common goal. At the beginning students work in groups of two to five allowing them to collaborate and dialogue as a way of sharing ideas, proposing more ideas, building consensus, and negotiating meaning in an effort to construct knowledge. Well-structured, challenging tasks lead to group engagement and active learning. Dialogue, presentations, writing and concept mapping prompt student groups to express their ideas and clarify their understanding. In this way, peer teaching/learning becomes a powerful classroom process.

Groups working together for a fairly long period of time develop into social groups that collaborate to make sense from the experiences of individual members; this is the essence of cooperative learning.

Reflection: The constructivist teacher is usually preoccupied with getting the students to engage in deep learning. The teacher attempts to get students engaged in organising their ideas into larger frameworks and making the connections between their ideas explicit, through activities such as concept mapping, semantic networking, writing, or making presentations. This is the essence of constructivist teaching that aims at promoting deep learning.

Safe Classrooms: Constructivist teaching promotes a safe environment in which students are treated with respect and encouraged to participate in thoughtful discussions. The teacher is not supposed to ridicule the students or use vulgar language against them or display some other non-respectful behaviour. The students are also encouraged to respect one another's ideas, raising questions and listening to each other. The teacher also provides in-teresting and challenging assignments to provoke critical thought. The classroom physical environment is also well arranged for learning. A good science classroom welcomes all students and strives to enable all motivated students to be successful.

Generally, a good classroom environment promotes student curiosity, rewards creativity, encourages a spirit of healthy questioning, avoids dogmatism, and promotes meaningful understanding.

Continuous Assessment: Teachers assess students' progress at each stage of the lesson and to identify the problems they face in understanding. The teachers also monitor the performance of groups, review journal comments, and ask probing questions. The teachers also assess students' progress at the end of each unit or term that reflect the kinds of activities and skills used in the lessons, and that focus on deep understanding of larger issues rather than small details. The teacher uses the outcomes to modify instruction as required.

Freedom and Privileging: The teacher needs to understand that scientific knowledge is produced by scientists who employ the scientific method. While the science that is taught in schools is selected from this original body of knowledge, the teacher does not have total freedom to teach all there is. The teacher essentially does privileging; some scientific knowledge is preferred over the others for the purpose of teaching. The differences between the two perspectives were clearly explained by Keys (1997). The teacher is expected to select carefully, with reference to the official curriculum, what the students ought to learn.

A Personal Instructional Philosophy in Biology

As a biology teacher, it helps a lot to develop a constructive instructional philosophy to guide your daily activities in an effort to achieve your career aspirations. An example of a personal instructional philosophy is shown in the textbox. What does the teacher believe is the purpose of teaching and how does this teacher hope to achieve the purpose? The teacher's instructional philosophy sounds like: *"Teach them well and let them lead the way"*

Box 7.1

Discussion

List as many items as you can regarding what effective constructivist biology teachers do.

Effective constructive biology teachers-------------

1. Create student - centred classrooms.
2. Create an environment in which students feel comfortable saying what they think.
3. Treat students with respect regardless of race, colour, gender, culture or other considerations.
4. Respect student ideas and treat them with care.
5. Become facilitators of students' learning rather than control agents or fonts of knowledge.

Every classroom presents a unique community of learners that varies not only in abilities but also in learning styles. My role as a teacher of biology is to give my students the tools with which to cultivate their own gardens of knowledge. To accomplish this goal, I will teach to the needs of each student so that all learners can feel capable and successful. I will present a curriculum that involves the interests of the students and makes learning relevant to life. I will incorporate themes, integrated units, projects, group work, individual work, and hands-on learning in order to make students active learners. Finally, I will tie the learning of biology into the world community to help students become caring and active members of society.

Activity:

Construct your own instructional philosophy that will guide you as a biology teacher.

Conclusion

The constructivist theory opines that learning is an active, constructive process by which the learners make of sense of their experiences and construct their own understanding of the content they study instead of having that understanding transmitted to them by someone else. Learners take an active role in building understanding and making sense of information. New information is linked to prior knowledge. For biology teachers to espouse this view of learning, they are expected to anchor their practice on constructivist instructional philosophy and to construct their own constructive philosophy of teaching biology.

References

Driver, R. Asoko, H., Leach, J., Mortimer, E. & Scott, P. (1994). Constructing scientific knowledge in the classroom. *Educational Researcher, 23* (7), 5-12.

Lijnse, P. L. (1995). "Developmental research" as a way to an empirically based "didactic structure" of science. *Science Education, 79*, 189-199.

Keys, Carolyn W. (1997) *'Perspectives on Inquiry-Oriented Teaching Practice: Clarification and Conflict'* a paper presented at the symposium at the Annual Meeting of the National Association for Research in Science Teaching at Chicago, April 21-24, 1997.

TEACHER SHOULD train students in laboratory skills, e.g. handling apparatus, measuring, making accurate observations, and in keeping records

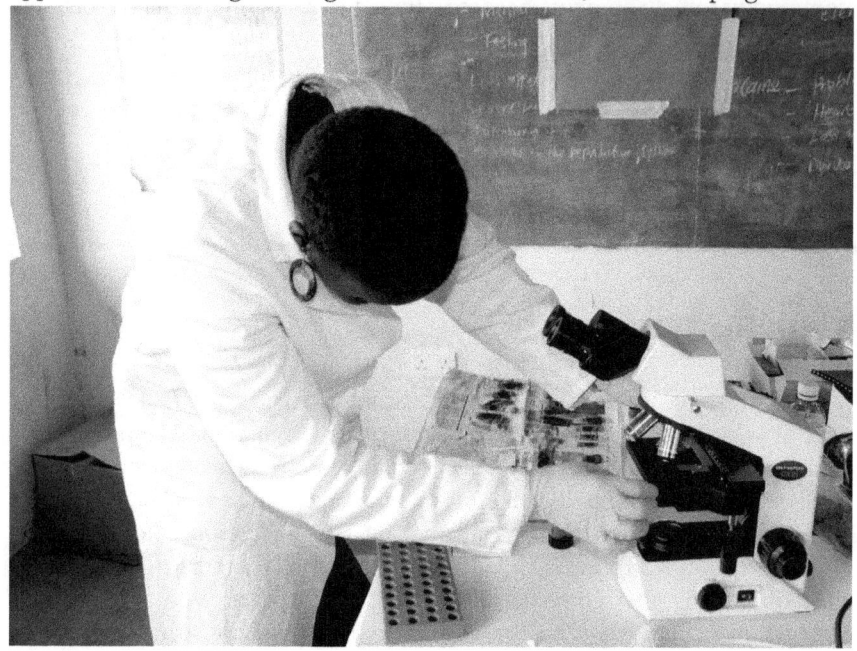

Photo: GLOBAL ALLIANCE FOR IMPROVED NUTRITION

CHAPTER EIGHT

PREPARING FOR INSTRUCTION IN BIOLOGY

◆ Setting up the Learning Situation ◆ The Unit Plan ◆ Schemes of Work

Introduction

One of the greatest challenges facing teachers of biology in schools is to find ways of meeting the learning needs of all the students in their class. Many teachers prepare lessons without considering important conditions of learning such as the knowledge, skills and experiences the students can bring to the lessons. In this process, teachers tend to over-rely on methods and approaches suggested in the biology teachers' guides. While it is very easy to use only the methods suggested in these guides, this is usually done at the expense of the very productive approaches.

In this chapter, you are encouraged to explore ways and means of creating an effective learning situation for the pupils. If the teacher can be more imaginative in the choice of lesson experiences, materials and approaches the biology classroom can become a more interesting place to be.

Setting up the Learning Situation

The biology teacher must plan carefully before committing students to any lesson in the classroom. The purpose of planning is to create a **learning situation** suitable for the students. The components of a learning situation (Figure 8.1) include **content, instructional objectives, instructional strategies, instructional resources**, and **evaluation**. The planning process focuses on each of these components to formulate the documents that guide the teacher in the classroom. The documents include the **Unit Plan, Schemes of Work,** and **Lesson Plan**.

The Unit Plan

A unit is a broad segment of subject matter having a common fabric of knowledge and therefore covers a biological theme or topic. A unit is organised around a specific topic so that it is neither a block of subject matter nor a series of independent lessons, but a careful organisation of subject matter and learning experiences. It usually takes from between two to four weeks of teaching. For example, the broad topic 'plants' is a unit covering different topics like seeds, germination, dispersion, classification, growth, reproduction, and transpiration. Unit planning is actually a middle ground between the scheme of work and lesson planning. It is longer than lesson planning but shorter than scheming planning. A well-planned unit integrates such subtopics and many types of learning activities.

Advantages of Unit Planning

- It breaks up a course into meaningful segments, small enough for pupils to easily grasp them. Most pupils work better on a series of short tasks than on a few large ones;
- It helps teachers to organise and present lessons in a cohesive, meaningful and logical manner;
- It helps the teacher to clarify the general and specific objectives of teaching;

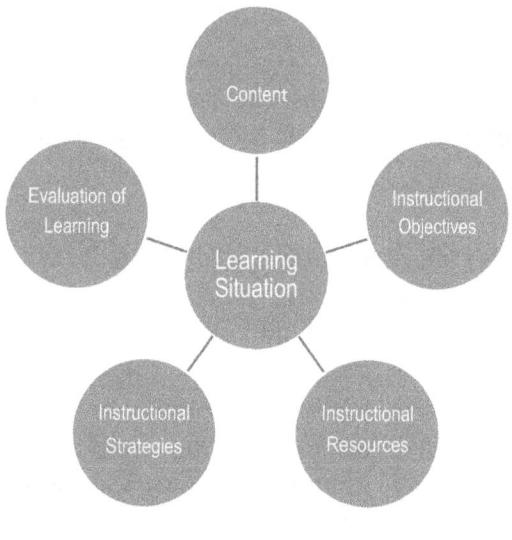

Figure 8.1: Components of a Learning Situation

- It helps the teacher to decide on the scope and sequence of subject matter as it relates to larger concepts;
- It helps the teacher to decide on the learning experiences and materials to be used in teaching a given topic;
- It helps the teacher to cater to the individual needs, interests and aptitudes of the learners; and
- It helps develop skills in the learners in a systematic manner, through planning several different kinds of activities.

Preparations before Making a Unit Plan

- Examine the Biology syllabus (school edition prepared by Kenya Institute of Education) and the Biology syllabus (examination edi-

tion prepared by Kenya National Examination Council) to determine the topic/areas to include in the unit plan;
- Examine records of work in the school to determine the entry behaviour of the students and to link the previous work to what must be covered in the unit;
- Identify the visual aids including demonstrations, texts and reference books available for use;
- Examine the laboratory setting to determine its usage during the course of the unit;
- Find out the usage of the library in case of homework/assignment requiring its use;
- Consider the variety of presentations to cover the aspects of the content for the unit, e.g. films, CD-ROM, field course excursions, etc;
- Consider how the interest and motivation of the learners could be sustained;
- Consider how students' participation in the learning process could be enhanced, e.g. use of laboratory activities, project work, etc;
- Consider ways of training students in laboratory skills, e.g. handling apparatus, measuring, making accurate observations, and in keeping records of what has been observed and what has been done e.g. drawing, making tables, etc; and
- Determining possible questions that may help students design observations or actively participate in learning and those that can be used for evaluating their learning.

Characteristics of a Good Unit

- The aims/objectives are clear and well defined;
- It caters for the needs, interests and capabilities of students;
- It provides for suitable learning experiences for students;
- It is flexible enough to provide for individual differences of the students;
- It includes inquiry-based instructional strategies like projects field course excursions, etc;
- It provides opportunities for students to relate biology to their lives and to other subjects;
- It provides a sequence that allows for progression and growth of students from year to year;

- The length of the unit should be such that it retains the interests and motivation of the students; and
- It provides adequate time for students to plan, organise and work on their own, and for the teacher to evaluate and follow up students' learning.

Steps in Unit Planning

Unit planning involves the following steps: Content analysis ('the what' of the unit); Objectives (the how of the unit); and Testing procedures (evidence of achievement).

Content Analysis

This involves examining the biology syllabus-both the school and the examination editions-to identify the content to be taught. The content is analysed in terms of facts, ideas, concepts, generalisations, situations, processes, laws, principles and theories. This helps in getting a thorough in-depth understanding of the subject matter (content) and in planning the lessons.

In content analysis, **concept maps, flow charts, classification charts,** and **concept trees/webs** are used to develop the ideas related to the concept or topic under investigation and in building the unit. Concept maps give better detail as they show the interconnectedness of concepts at similar hierarchical levels, whereas the others are linear, progressing from one point to another without showing the linkage between branches. For example, in Figure 8.2, the concept map of respiration is constructed according to the guidelines indicated next.

Procedures for Constructing Concept Map

- Select the idea/concept to form the top base;
- Generate a set of concepts/ideas associated with the main topic or concept;
- Rank the concepts/ideas hierarchically from most general (or most inclusive) to most specific (or least inclusive). Group the concepts/ideas that are related.
- Draw the concept map with the concepts in ovals/boxes as follows:
    o Most general concepts at the top;
    o Intermediate concepts below;

- o Most specific concepts at the bottom; and
- o More general concepts normally connected to two or more specific concepts.
- Draw the lines connecting the concepts;
- Write in the words describing the relationships;
- Draw in the lines;
- Write in the linking words; and
- Revise if necessary.

Objectives

The teacher should identify the general and specific objectives of the unit, after going through the content and selecting the desirable ideas, facts, generalisations, theories, laws and issues or situations.

Learning Activities

The teacher selects classroom/laboratory activities such as laboratory investigation tasks, and questions to be investigated; instructional strategies to be used in teaching the content such as demonstrations, experiments and fieldwork; and resources to be used such as texts, references, films, slides, and library research.

The activities should be designed keeping in mind individual differences, the psychology of learning, the content and objectives. These should stimulate students' thinking and arouse their curiosity.

Testing Procedures

The teacher should then select suitable evaluation tools and techniques to test students' understanding of the content to be covered and the effectiveness of the teaching strategies to be used. The teacher in this process should examine the school internal tests and past examination papers to determine the nature and scope of examination on the content to be covered in the unit.

Example of Unit Plan

Subject: **Biology** Class: **Form 2** Term: **3** Name of Unit: **Respiration**

Number of Lessons: **16 (4 weeks)**

The concept map for respiration is developed according to the procedure discussed above for content analysis. The resultant concepts and sub-concepts are indicated in Figure 8.2. The content of the unit is then drawn from the concept map and displayed as displayed below.

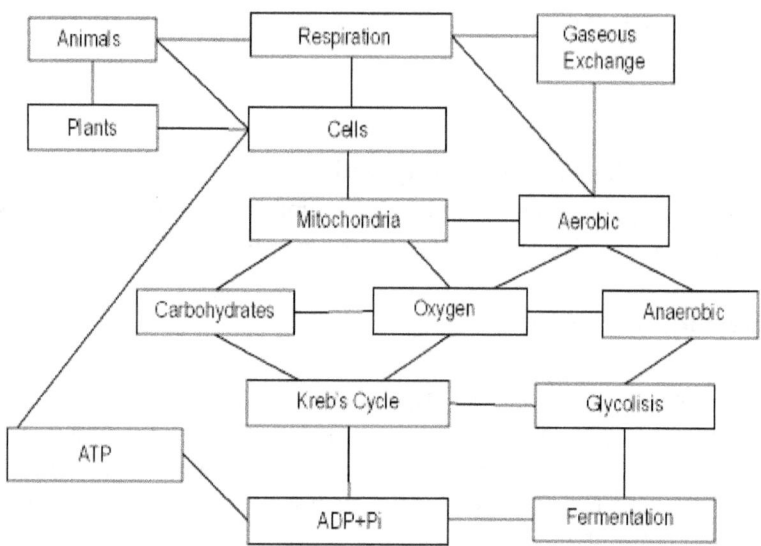

Figure 8.2: The Concept map of Respiration

1) Content Analysis
a) Meaning and Significance of Respiration;

- Products of respiration;
- Respiration and gaseous exchange; and
- Types of respiration.

b) Tissue respiration;

c) Structure and Function of Mitochondrion;

d) Aerobic Respiration;

- Aerobic respiration in plants and animals.

e) Anaerobic respiration;

f) Application of Anaerobic respiration in the industry and at home; and

g) Comparison of Energy output of Aerobic and Anaerobic Respiration.

2) Specific Objectives

By the end of the unit learners should be able to:

a) Explain the meaning and significance of respiration in living organisms;
b) Investigate the outputs of respiration in living cells such as germinating seeds;
c) Distinguish between aerobic and anaerobic respiration;
d) Explain the process of respiration in plants and animals;
e) State the factors affecting the rate of respiration; and
f) Discuss the applications of anaerobic respiration in industries and in homes.

3) Learning Activities

Group work and independent study to:

a) Investigate the gases that are given off when food is burnt;
b) Investigate production of heat energy by germinating seeds;
c) Investigate the gas production during fermentation;
d) Discuss the differences between aerobic and anaerobic respiration;
e) Determine the respiratory Quotient of the various respiratory substrates;
f) Investigate the factors affecting the rate of respiration; and
g) Describe the structure of mitochondrion and explain how it's suited for its functions.

4) Evaluation
a) Solve problems at end of the chapter in the class textbook;
b) Classroom test framed with questions or content in the unit; and
c) Laboratory exercises to evaluate whether the student can find the outputs of respiration in plants.

Schemes of Work

Definition of Scheme of Work

A scheme of work is a breakdown of work to be taught to a specified class over a period of time, usually one term. It is usually compiled from the unit plans.

Things to do before Scheming:

- Examine the Biology syllabus (school edition prepared by Kenya Institute of Education) and the Biology syllabus (examination edition prepared by Kenya National Examination Council) to determine the topic/areas to include in the unit plan and schemes of work. **A syllabus is an outline of the content to be taught and/or examined to a specified group of students for a specified period of time;**
- Find out time allocation for the topics in the syllabus (weighting) and for biology course on the school time table and estimate the time needed to cover the work planned for;
- Examine records of work in the school to determine the entry behaviour of the students and to link the previous work to the work be covered in the unit, i.e. to find out if the students already have background knowledge and how much so that the required time is allocated;
- Examine pupils' textbooks and library reference books and determine the reading task, assignments or exercises;
- Identify the visual aids including demonstrations, texts and reference books available for use;
- Examine the laboratory setting to determine its usage during the course of the unit;
- Consider the variety of presentations to cover the aspects of the content for the unit, e.g. films, CD-ROM, field course excursions, etc;
- Consider how the interest and motivation of the learners could be sustained;
- Consider how students' participation in the learning process could be enhanced, e.g. use of laboratory activities, project work, etc.;
- Consider ways of training students in laboratory skills, e.g. handling apparatus, measuring, making accurate observations, and in keeping records of what has been observed and what has been done e.g. drawing, making tables, etc.;
- Determining possible questions that may help students design observations or actively participate in learning and those that can be used for evaluating their learning. Plan the testing procedure to determine the success of teaching and learning. Check on the past examination papers both external and internal to help in constructing the questions; and

- Decide on the most appropriate mode of communication: Oral story by the teacher; Handouts; Worksheets; Discovery task; Visual aids; and Practical work.

Advantages of Scheming

- Takes into account the level from which the pupils will start and provides effective direction;
- Breaks up the course into meaningful segments small enough for the pupils to grasp easily;
- Helps the teacher to organise and present lessons in a cohesive, meaningful and logical manner;
- Helps the teacher to plan ahead for the learning experiences for a given topic such as expeditions, practical, excursions, discussions, projects, etc.;
- Helps the teacher to realise the need for use of a variety of teaching approaches and resources and to plan for them; and
- Helps the teacher to decide on the scope and sequence of subject matter and to distribute them according to their weighting as it relates to the larger concepts. Major topics require more time than smaller topics.

The format for Schemes of Work

A good scheme of work should contain:

- Time allocation, in terms of weeks and terms.
- Statement of content in terms of topic, subtopic and subject matter
- Aims and objectives
- Teaching resources
- Learning activities including teaching approaches
- Evaluation/remarks

Table 8.1: Format for the Scheme of Work

Week	Lesson	Topic	Content	Objectives	Teaching/ Learning Activities	Learning resources	Remarks/ evaluation

A Sample Scheme of Work

SUBJECT: Biology

CLASS: Form 2 Red   **TERM:** 2

NAME OF TEACHER: Mr Onyango

Table 8.2: A Sample Scheme of Work in Biology

Week	Lesson	Topic	Content	Objectives	Teaching and Learning Activities	Learning Resources
I	1 - 2	Respiration in plants and Animals	-Meaning and significance of respiration -Tissue respiration and gaseous exchange	By the end of the lesson pupils should be able to: - Define respiration. -Explain the significance of respiration -Distinguish tissue respiration from gaseous exchange	- Class discussion to define and explain the significance of respiration. -Discussing the difference between tissue respiration and gaseous exchange -Demonstration to illustrate output of burning food in oxygen	Secondary Biology Students' BK 2., P. 85 -Starch powder, crushed beans, maize flour, milk powder, boiling tubes, test-tubes, calcium hydroxide solution, rubber stopper, anhydrous cobalt chloride paper, Bunsen burner, delivery tubing and retort stand

	3		-Types of respiration -Structure and Function of a mitochondrion -The overall process of respiration	-Distinguish between aerobic and anaerobic respiration -Draw and label the external and internal features of a mitochondrion - Describe the overall process of tissue respiration	-Class discussion to distinguish between aerobic and anaerobic respiration -Drawing and labelling internal and external structures of mitochondrion -Class discussion to describe the overall process of respiration	Secondary Biology Students' BK 2. P. 86
	4		Aerobic respiration	-Explain the stages of glycolysis and Kreb's cycle in aerobic respiration -Summarise the equations of aerobic respiration in plants and animals	-Class discussion to explain the stages in the aerobic respiration in plants and animals -Complete the equations of aerobic respiration in plants and animals	Secondary Biology Students' BK 2. P. 86-88
II	1-2		Aerobic respiration	-Plan and conduct experiment to determine the production of heat by germinating bean seeds	-Students plan and conduct an experiment to find out production of heat energy in germinating bean seeds	-Boiled bean seeds, soaked bean seeds, vacuum flasks, cotton wool, thermom-

					-Answering practical activity tasks in the pupil's book page 88-89	eters, methanol, retort stands.
	3 - 4		Anaerobic respiration	-Explain the stages of anaerobic respiration in plants and animals -Summarise the equations of anaerobic respiration in plants and animals -Plan and conduct experiment to determine the production of gas by yeast in a sugar solution	- Class discussion to explain the stages in anaerobic respiration in plants and animals -Draw out the equations of anaerobic respiration in plants and animals -Students plan and conduct an experiment to find out gas production by yeast in a sugar solution -Answering practical activity tasks in the pupil's book page 91	Secondary Biology Students' BK 2. P. 89-91 -Boiling tubes, measuring cylinders, test tubes, thermometers, delivery tubes, rubber stoppers, 10%glucose solution, yeast, oil, retort stands, bunsen burners.
III	1 - 2		Anaerobic respiration	-Explain the role of anaerobic respiration in oxygen debt.	-Class discussion on oxygen debt and the role of anaerobic respiration	Secondary Biology Students' BK 2. P. 89-91

					in this.	
					-Collating and interpreting data on experiments conducted on anaerobic respiration.	
	3		Comparing Aerobic and Anaerobic respiration	-Determine the similarities and differences between aerobic and anaerobic respiration	-Class discussion on similarities and differences between aerobic and anaerobic respiration	Secondary Biology Students' BK 2. P. - 92
	4		Respiratory Substrate	-Analyse the difference in energy production of various respiratory substrates	-Groups rank amount of energy provided from various respiratory substrates -Collating and interpreting data on experiments conducted on aerobic respiration	-Charts containing data on the amount of energy provided by various respiratory substrates
IV	1 - 2		Respiratory Quotient	-Explaining the meaning and significance of respiratory quotient -Calculate respiratory quotient of various respi-	-Class discussion to define and determine the significance of RQ -Groups work out RQ of various	-Charts containing data on the volume of carbon dioxide and oxygen produced by

				ration activities	respiratory activities	various respiratory activities Secondary Biology Students' BK 2. P. 93
3			Factors affecting the rate of respiration	-Explaining the factors affecting the rate of respiration	-Class discussion to explain factors affecting rate of respiration	Secondary Biology Students' BK 2. P. 93-94
4				-Applying the knowledge gained on respiration	-Individual work on answering revision questions	Secondary Biology Students' BK 2. P. 95-96

Conclusion

Preparing to teach is a significant stage in the biology instructional process. Many biology teachers prepare lessons without considering important conditions of learning such as the knowledge, skills and experiences the students can bring to the lessons. The teachers tend to over-rely on methods and approaches suggested in the biology teachers' guides. While it is very easy to use only the methods suggested in these guides, this is usually done at the expense of the other very useful instructional strategies. This chapter has guided the teacher on how to explore ways and means of creating an effective learning situation for the pupils in terms of choice of lesson learning experiences, materials and instructional approaches. This means that the biology teacher must plan carefully before committing students to the lesson. The purpose of planning is to create a **Learning Situation** suitable for the students. The most basic product of the preparation process is the design and construction of the schemes of work which is predicated on the choice of lesson learning experiences, materials and instructional approaches covering a long period of time usually one school term.

References

Lemlech, J. Kasin (2010). *Curriculum and Instructional Methods for Elementary and Middle School.* 7th Edn. Boston: Allyn and Bacon

Farrant, J (1980). *Principles and Practice of Education.* Lagos: Longman

LESSON PLANNING is one of the most important aspects of planning to teach Biology

Photo: STEVE REEKIE/FLICKR

9

CHAPTER NINE

LESSON PLANNING IN BIOLOGY

◆ Definition of Lesson Plan ◆ Thing to be done before Lesson Planning ◆ Characteristics of a Good Lesson Plan ◆ Procedures of Lesson Planning for Teaching Biology ◆ Stating Instructional Objectives in Biology ◆ Designing the Lesson Plan for Teaching Biology ◆ Lesson Plan Format ◆ A Specimen Lesson Plan in Biology

Introduction

The teacher of biology must plan carefully before committing students to any lesson in the classroom. The purpose of planning is to create a **Learning Situation** suitable for maximum engagement of the students in the lesson. The students' engagement in the lesson will depend on how well the teacher selects the **content** and constructs the **instructional objectives on one hand and the instructional strategies and resources and evaluation** procedures on the other. The lesson planning process focuses on each of these components to formulate the lesson plan that effectively guides the teacher in the classroom. This chapter outlines the mechanisms of designing and constructing a lesson plan for the effective teaching of biology.

Definition of Lesson Plan

A lesson plan is a detailed daily overview of what the teacher intends to do with the learners in terms of **what they should learn, how it will be presented, and how it will be established if they have learned.** It outlines the important aspects of a lesson arranged in the order in which they are to be presented to the learners. **A lesson plan is, therefore, a teacher's plan of action.**

Things to be done before Lesson Planning

Lesson planning is one of the most important aspects of planning to teach biology. A lot of thinking goes into it. However, it becomes easier for the teacher if both the schemes of work and the unit plan have already been prepared. From the unit plan, the teacher will just be required to select a problem or set of problems, the objectives and the instructional resources and strategies for a given lesson. Before writing the lesson plan it is, therefore, necessary to carry out the following:

- Collect information on the topic you wish to teach, using a variety of resources. The information includes: concepts, diagrams, practical activities, demonstrations, problems to get students investigating, end-of-lesson activities such as suggested homework, materials required such as worksheets, apparatus, equipment, and visual aids;
- Decide on the exact content of the particular lesson based on the time available. If the topic is wide, select and sequences the subtopics as they should appear in your teaching. The topic may take more than one or two lessons; and

- Decide how you will teach the lesson:

    o The specific objectives of the lesson - i.e. the purpose of communicating the proposed knowledge, skills and attitudes to be developed;
    o Introduction to the lesson i.e. motivation, linkage, recall, etc.;
    o The learning activities the learners will be engaged in and the time required for them;
    o The organisation of the learning activities;
    o The teaching strategies and the learning resources to be used;
    o The type of records to be kept by the learners;
    o The end activities of the lesson; and
    o Evaluation of the success of the lesson.

Advantages of Planning a Lesson

The following are important advantages of making and using the lesson plan for the teaching of biology:

- The objectives of the lesson are made clear both to the teacher and the learner;
- Permits the teacher to anticipate pupils' problems and reactions so as to prepare adequately in order to avoid foreseeable difficulties;
- Makes the teacher's presentation well organised and systematic;
- Enhances the teacher's self-confidence and self-reliance in conducting the lesson;
- Helps the teacher to manage time effectively as every step is planned with forethought and repetition is avoided;
- Facilitates the choice of appropriate teaching strategies and resources;
- Establishes a proper connection between different lessons, thus ensuring continuity in the teaching and learning process;
- Facilitate retention of students' interest in learning through the use of suitable activities and assignments;
- Permits the teacher to plan for individual differences among the learners according to their mental level and gender;

- The main points of the lesson are concisely summarised in a systematic manner during the lesson and at the end of the lesson; and
- Review of the lesson is done within the allocated time.

Characteristics of a Good Lesson Plan

A good lesson plan is that which bears the following characteristics:

- Objectives are very clear and achievable - i.e. SMART (Specific, Measurable, Achievable, Realistic, and Time-bound);
- It can be used by another teacher and still achieve the intended objectives;
- It is well prepared;
- The content selected takes into account the nature of the subject, past experiences of the learner and the requirements of the syllabus;
- The teaching strategies and resources planned to take into account the objectives, and the needs of the learners;
- It ensures active participation of the learners through suitable learning activities;
- Motivates the learners through the use of stimulating resource and materials and teaching strategies;
- Ensures use of good, well-thought-out questions in developing and reviewing the lesson within the allotted time;
- Ensures effective and sufficient use of chalkboard for presentation and summarising; and
- Ensures the main points are concisely summarised systematically.

Procedures of Lesson Planning in Biology

The procedures that enable you to design and construct an effective lesson plan include the following:

- Identifying the overall lesson topic/title;
- Determining the goals/ or objectives of the lesson (instructional objectives);
- Designing the lesson plan;
- Determining the materials & supplies needed (instructional resources); and

- Constructing the sample outline for a Lesson Plan.

Stating Instructional Objectives in Biology

The instructional objectives in biology are selected and constructed systematically by addressing the following dimensions:

a) What is an Instructional Objective?

An instructional objective is a statement of intent that specifies, in behavioural terms, the outcome of a learning activity. An objective indicates:

- Behaviour to be achieved (knowledge, skills, attitudes);
- Standard to attain;
- Condition; and
- Audience (pupil-centred).

b) Examples of Instructional Objectives

Incorrectly stated:

The objective of this lesson is to teach the equation summarising the process of respiration and let the learner understand how it is used to calculate the amount of heat energy produced.

Why is this Objective unsuitable?

This instructional objective is not correctly stated due to the following reasons:

- The behaviour is not well articulated – the terms "teaching" and "understand" are neither behavioural nor measurable;
- A formula or equation is not taught in isolation; we must provide appropriate activities for learners to apply the formula to familiar and unfamiliar situations;
- The condition is not indicated. We are not sure at what point in time is the objective to be achieved; and
- The audience is not addressed. The objective is addressing the teacher (teacher – centred) rather than the learner.

Correctly Stated:

The objective should be correctly stated in co-operating the components of an instructional objective as follows:

- By the end of the lesson, the learner should be able to calculate the respiratory quotient; and
- By the end of the lesson, the learner should be able to correctly calculate all problems relating to the respiratory quotient.

The first objective does not contain the standard, while the second one has the standard. The standard indicates the level at which the teacher would be satisfied that the students have performed adequately. Quite often many teachers erroneously ignore this level of achievement.

c) Commonly used Action Verbs in the construction of Objectives

In stating instructional objectives in behavioural terms the teacher must select action verbs appropriately basing on Bloom's cognitive levels. Table 9.1 shows some of the action verbs commonly used in the construction of instructional objectives in biology.

Table 9.1: A List of Action Verbs used in Constructing Lesson Plans

COGNITIVE LEVEL	EXAMPLE OF ACTION VERB			
Knowledge/Recall	Define State Name	List Write Enumerate	Underline Label Select	Reproduce Measure Recall
Comprehension	Identify Indicate Illustrate Translate	Formulate Explain Describe Interpret	Classify Elaborate Compare	Contrast Represent Justify
Application	Predict Assess Explain Perform	Find Show Demonstrate Construct	Calculate Compute Use Determine	Interpret Solve Design Equate
Analysis	Analyse Conclude Differentiate	Elucidate Separate Resolve	Intergrate Criticise Judge	Account Outline Trace
Synthesis	Combine Restate Summarise Verify	Argue Discuss Organise Clarify	Derive Relate Generalise	Deduce Conclude Infer

Evaluation	Evaluate	Critique	Propose	
	Recognise	Support	Attack	
	Defend	Adduce		

d) Indicators for Cognitive Levels for the Objectives

Each cognitive level is characterised by indicators, some of which are shown in Table 9.2.

Table 9.2: Indicators of Cognitive Levels used in Constructing Objectives

Cognitive Level	Indicators
Knowledge/Recall	- Recall of facts/specific information - Characterised by recognising, memorising, and remembering
Comprehension	- Ability to transform information - Characterised by interpreting, translating, describing in own words
Application	- Using the information to solve new problems - Characterised by solving problems
Analysis	- Ability to separate the whole into its components - Characterised by subdividing, taking apart thoughts and problems
Evaluation	- Ability to suggest well-reasoned decisions on debatable ideas - Characterised by resolving differences of opinion

Designing the Lesson Plan for Teaching Biology

An effective lesson plan is constructed by seeking answers to the following guiding questions:

Preliminary information

- What will be the topic of the lesson?
- What grade level shall you teach?
- How much time shall you need?

Introduction to the Lesson

- What will you say to help students understand the purpose of the lesson?
- How will you help them make connections to prior lessons or experiences?
- How will you motivate them to become engaged in the lesson?

Lesson Development

- During the lesson, what are the specific details of activities about how you will begin and end the lesson?
- What teaching and learning supplies and materials will you need for the lesson?
- What discussion questions will use?
- How will you help children understand behaviour expectations during the lesson?
- When/how you will distribute supplies and materials?
- How much time will you require for the lesson development?

Summary of the Lesson

- How will you bring closure to the lesson and help children reflect on their experiences?
- How will you help them make connections to prior lessons or prepare for future experiences?
- What kind of feedback do you want from them at this time?

Assessment

- What academic, social, and linguistic support will you give the students during the assessment?

Lesson Plan Format

Consistent with the design, the lesson plan is constructed based on the general components of a good lesson plan:

- Title (Administrative details);
- Instructional objectives;
- Statement of teaching resources (including resources in references);
- Subject matter (content);
- Time utilisation (steps/stages);
- Introduction (set induction);
- Lesson development;
- Conclusion (closure);

- Self-evaluation; and
- Lesson notes.

The components are next described while the lesson plan format is summarised in Table 9.3.

Table 9.3: Format for Lesson Plan in Biology

TIME/STEPS	CONTENT	TEACHER'S ACTIVITIES	PUPILS' ACTIVITIES	INSTRUCTIONAL RESOURCES	REMARKS/ SELF EVALUATION

Title/ Administrative Details

NAME OF TEACHER: --

CLASS: -------------------------NUMBER OF PUPILS IN CLASS: ---------------------

SUBJECT: --

TOPIC: --

SUBTOPIC: ---

TIME: ------------------------------------DATE: -------------------------------------

Instructional Objectives

References

Description of the Components of the Biology Lesson Plan

Title/Administrative Details

This states the name of the teacher, the subject the teacher plans to teach, when and **to whom** the lesson is to be taught, the topic and subtopic to be taught and the number of learners to attend the lesson. This information assists the teacher in selecting the teaching approaches and resources to use.

Instructional Objectives

The objectives cover all the main achievements planned for in the lesson and correspond to those stated in the schemes of work. The objectives should be realistic, not too ambitious and not too limited. The objectives should be: stated in behavioural terms focusing on specific knowledge, skills and attitudes to be developed in the learner; pupil-centred, and stated with regard to the level/criteria of expected learner behaviour (the standard).

Learning Activities

These are the activities the learner shall be engaged in to gain proficiency in knowledge, skills and attitudes planned for. The activities may include laboratory practical work or class experiments; discussions; field course excursions; reading and making notes; project work; listening to guest speakers, among others.

The learning activities should be as varied as possible to include student/teacher controlled activities; group/individual activities; use of all the senses; and use of full range of multimedia. The activities should encourage full participation by the learners and should be interesting and challenging enough to arouse and sustain learner's interest.

Learning Resources

Resources refer to teaching aids as well as references to be used. There should be a logical connection between the learning activities and the resources that support them throughout all the stages of the lesson. Good resources should be durable for future use, clear to the whole class, attractive, cheap and easily improvised.

Content

The content is directly suggested in the syllabus but is organised in a sequence suitable for the learner in the lesson plan. Content is written in a brief form indicating what the learners will learn, but details are included in the separate **lesson notes**. For this purpose, the content is planned to be presented in three major parts: **Introduction; lesson development;** and **conclusion.**

Introduction (Set Induction)

Introductory activities are initiating tasks of the lesson intended to prepare the learners for the lesson. The functions of these activities are to:

- Gather learners' attention and focus it on the lesson;
- Help the teacher to check on the relevant prior knowledge of the learners concerning the topic (i.e. entry behaviour); and
- Provide a link between the previous knowledge and the current lesson.

The activities may include:

- Testing of previous knowledge and degree of receptivity; and
- Arousing curiosity and maintaining inclination by means of activities like telling a story, demonstration, displaying a chart, showing a model, giving a puzzle or problem, etc.

Lesson Development or Presentation

Lesson development refers to the central or main tasks the pupils will be involved in and the teacher's instructional strategies and materials to be used in order to achieve the results. While the activities should be learner-focused both the teacher and the pupils should be active participants in this enterprise.

In the lesson development, the teacher systematically and logically develops the lesson by professionally guiding the pupils' learning using a variety of techniques like questioning, demonstrations, lectures, discussions, and practical work in a predetermined sequence and thus delivering the content and enabling them to develop the necessary skills and attitudes. The teacher may reiterate definitions of required terms to familiarise pupils with the best of them, provide occasional explanations, summarise and restate, give alternative explanations for phenomena, or the teacher may guide pupils to make comparisons and associations, generalisations, and applications to new situations to enhance learning. These techniques help the teacher to make the lesson more interesting, comprehensive and meaningful. The teacher also uses various instructional resources such as charts, models, textbooks, apparatus, equipment, and references.

The content is usually divided into about three or more teachable but hierarchical steps. Each step should contain what the learners are expected to know or do at its completion and should be directly linked to the learner's actual classroom behaviour (learning objectives). The teacher is expected to observe strict timing for each activity in each step.

The chalkboard summary should be developed alongside the lesson development. The teacher in the entire process refers to the lesson notes. The notes should be brief but adequate to efficiently guide the teacher through the lesson. The notes may be kept separate from the details of the lesson plan, or they may be integrated into the lesson plan, depending on the teacher's experience and background. If the teacher has problems in working with lesson notes and lesson plan separately, the paper may be divided into two so that at a glance the activities students are supposed to carry out and what the teacher is supposed to do are seen.

Conclusion or Closure

The conclusion involves culminating activities that put together the main aspects of the lesson. The conclusion is important because it **allows learners to put the work into a proper perspective and for recapitulation and transference of content that has just been learned. It also helps the teacher to judge how well the pupils have learned the content and how effective the teaching strategies have been.**

The functions of closure are to:

- Help draw students' attention to a closing point in the lesson;
- Help students consolidate their learning – i.e. tying all the information or activities into a meaningful whole;
- Reinforce the major points to be learned so that the students are helped to retain the important information learned in the lesson;
- Relate the lesson to the original organising concept or principle or theme;

- Lead students to extend or develop new knowledge from previously learned concept or principle; and
- Allow students to practice what they have learned.

The lesson can be concluded through a variety of activities, including:

- Taking notes, for example in the form of filling in the blanks;
- Oral questioning on main aspects of the lesson;
- Asking pupils to label sketches or draining;
- Organise thinking around a new concept or principles;
- Follow – up to the film, T.V programme or Video show;
- Summarising the main points of the lesson on the chalkboard;
- Providing information for the next lesson (advance organiser)-sign post what shall be covered in the next lesson;
- Giving an assignment to consolidate the content that has been learnt so far, by providing sources of information, and objectives and format of the assignment report;
- Administering a brief evaluation exercise to ascertain whether the pupils have understood the content taught by asking suitable questions on the lesson orally, giving a short objective type test, asking them to label sketches or fill in the blanks, and giving an assignment on related aspects of the lesson; and
- Giving a project focusing on designing a product or an approach to solving a problem, testing a hypothesis, locating and using information on their own, modifying an old or existing product.

Lesson Notes

The lesson notes prepared by the biology teacher should be brief but providing adequate information for the students on the given topic or concept. The notes are preferably prepared in a separate notebook from the lesson plan. In preparing the notes the teacher should look at the past examination papers so that the notes address the levels of knowledge required to be attained by the students.

Example 1: Lesson Notes

Subject: Biology

Class: Form 4

Topic: The Chromosome

Sub-topic: Genes and DNA

Teaching Points:

<u>Structure and replication of DNA</u>

- Chromosomes are thread-like structures found in the nuclei of cells. They contain the genetic materials, the Deoxyribonucleic acid (DNA);
- DNA is the genetic material that is transmitted from parent to offspring. These factors are known as genes;
- DNA is a nucleic acid made of two strands of nucleotides wound together in a spiral called a double helix. Each nucleotide is composed of a sugar molecule known as deoxyribose, a phosphate group, and one of four different nitrogenous bases: adenine (**A**), thymine (**T**), guanine (**G**) or cytosine (**C**);
- The phosphate and sugar parts of the nucleotides form the backbone of each strand in the DNA double helix;
- The bases extend toward the centre of the double helix, and each base in one strand is matched with a complementary base in the other strand, in accord with the base-pairing rules: **A** pairs with **T** and **G** pairs with **C**;
- The DNA is therefore like a twisted ladder with the nitrogen bases forming the steps or rungs;
- These characteristics are the same for the DNA of all organisms. The DNA of different organisms differs in the sequence of nucleotides, and these differences in nucleotide sequence are responsible for the genetic differences between different organisms;
- DNA replication produces two new DNA molecules that are identical to the original. DNA molecule, so each of the new DNA molecules carries the same genetic information as the original DNA molecule;
- During DNA replication, the two strands of the original DNA double helix are separated and each old strand is used as a template to form a new DNA strand. The enzyme DNA polymerase adds nucleotides one-at-a-time, using the base-pairing rules to match each nucleotide in the old DNA strand with a complementary nucleotide in the new DNA strand. Thus, each new DNA double helix contains one strand from the original DNA molecule, together with a newly synthesised matching DNA strand; and

- In eukaryotic cells, each chromosome consists of DNA wrapped around proteins. The chromosomes are contained in the nucleus inside a nuclear membrane.

Students Worksheet

Worksheets are usually a series of exercises or practical tasks for the students, mostly produced by the teacher. They may also include summarised notes. The tasks usually require students to make decisions, give reasons for observations conducted, evaluate strategies used and construct hypotheses that spur inquiry (Petty, 2009).

<u>**An Example**</u>

Subject: Biology

Name of Student-------------------------------------**Class**----------------

Topic: Investigating Chromosome Behaviour in Mitosis

Introduction:

Chromosomes are thread-like structures found in the nuclei of cells. Along the length of the chromosomes is a series of structures called genes. The genes contain the genetic materials, the Deoxyribonucleic acid (DNA). Chromosomes exhibit characteristic behaviour during cell division, which ensures the transmission of genetic material from parent to offspring.

Equipment and Supplies:

- Thick brightly coloured thread
- A pair of scissors
- White Manila paper
- Transparent cellotape
- Pictures of a random arrangement of homologous chromosomes at metaphase I (independent assortment)

Procedure

1) Cut 18 pieces of thread each 6cm long. Make another set of 18 threads 3cm long. These pieces of thread represent chromo-

somes.
2) Study the stages of mitosis and use them as a guide to make model stages of mitosis. Assume the cell has two pairs of chromosomes.
3) Draw the cell outline for each stage on the manila paper. Based on the information, identify each stage of mitosis and put the heading on each outline.
4) Use the threads to make models of chromosomes. For the centromere, tie a knot on the threads.
5) Stick the chromosomes in position using cellotape. Using a pencil, draw the centrioles and spindle fibres.

Questions:

1) How important is the behaviour of chromosomes during cell division?
2) Comment on the chromosomal number of parent cell and that of daughter cells

A Specimen Lesson Plan in Biology

NAME OF TEACHER: --

NAME OF SCHOOL: Kakamega High School

CLASS: Form 1 R   **NUMBER OF STUDENTS:**   50

SUBJECT: Biology   **DATE:** 5th May 2018 **TIME:** 8.00 – 9.20

TOPIC: Classification I

SUB-TOPIC: Scientific Naming of Organisms

OBJECTIVES

By the end of the lesson, pupils should be able to:

(a) Discuss the necessity and significance of classification of organisms;
(b) Outline the major units of classification of organisms; and
(c) Apply Binomial nomenclature in classifying organisms.

Table 9.4: A Sample of Lesson Plan

TIME & STEP	CONTENT	TEACHER'S ACTIVITIES	PUPILS' ACTIVITIES	INSTRUCTIONAL RESOURCES	REMARKS
Introduction (5 minutes)	Introduction to classification of organisms	- Tell a story of a herbalist who visited a forest in the neighbouring country that threw him into conflict with local herbalists over the names and functions of several plants in that forest. - Establish by discussion that in fact the conflicting names of the plants was unnecessary as the common names differed from place to place and the need to use scientifically established ones. - Instruct pupils to pair up and move to where apparatus are	- Pupils respond to the excursion story by giving their experiences regarding use of names - Pupils discuss the need for use of scientific names -Pupils form groups of two for each working station	Secondary Biology, Form 1 P.14-19	-Story telling and whole school instructional techniques used

Table 9.5: A Sample of Lesson Development Plan

Lesson Development					
Lesson Development	Classification of some organisms	- Introduce pupils to the charts showing 7 units of classification of some common organisms such	- Pupils check the charts and familiarise with the 7 units of clas-	-Charts of classification - Worksheets	-Questioning and whole class instructional techniques used

Step					
Step I (20 minutes)	-Definition of binomial nomenclature -Need and significance of scientific naming of organisms.	as humans, dog, cat, maize and bean. -Asking pupils to define binomial nomenclature.	sification. -students define binomial nomenclature -Students discuss need and significance of binomial nomenclature		
Step II (20 minutes)	Scientific naming of organisms	- Asking pupils to read worksheets in groups and write scientific names of specimens provided. - Asking students to read in groups and answer questions in relevant the worksheet	-Students read in groups instructions in worksheets and provide answers on binomial nomenclature	-Worksheets -Previously collected plant and animal specimens -Preserved plant and animal specimens -Pictures of various organisms	-Group discussion used
Step III (20 minutes)	Scientific naming of organisms	- Conducting whole class discussion of pupils findings -Ask students to complete the classification table given in the worksheets	-Students give their findings and discuss them -Students complete the classification table in the worksheet	Pupils textbook page 18	-Group discussion used
Conclusion (10 minutes)	Summary of lesson	-Using the summary chart to review with whole class through oral questioning. - Emphasise the key points	- Pupils provide answers to questions and take down summary notes from CB.	- Chart	

Conclusion

This chapter has outlined the important steps in the design and construction of the lesson plan for the teaching of biology. The lesson plan is predicated on the schemes of work developed by the teacher as outlined in the previous chapter. The students' active engagement in the lesson will depend on how well the teacher selects the content, constructs the instructional objectives and selects the instructional strategies, resources and evaluation procedures. The lesson planning process focuses on each of these components to formulate the lesson plan that effectively guides the teacher in the classroom.

References

Farrant, J (1980). *Principles and Practice of Education*. Lagos: Longman

Lemlech, J. Kasin (2010). *Curriculum and Instructional Methods for Elementary and Middle School*.7th Edn. Boston: Allyn and Bacon

Petty, G (2009). *Teaching Today. A Practical Guide*.4th Edn. Chelternum: Nelson Thornes

LEARNING IS more effective if students are actively involved in obtaining the information

Photo: USAID/CAROLE DOUGLIS/FLICKR

10

CHAPTER TEN
METHODS, TECHNIQUES AND STRATEGIES OF TEACHING BIOLOGY

◆ Teaching Activities ◆ Meaning of Teaching Method, Technique and Strategy ◆ Factors Influencing choice of Teaching Strategies in Biology ◆ Transmission Method Techniques ◆ Problem Solving Method

Introduction

In the planning process, it is important to consider the best teaching and learning activities that can be used to realise the instructional objectives. The activities should be communicated to the learners in the manner suitable to them so as to cater to their interests and needs and thus facilitate the learning process. This chapter outlines some of the commonly used instructional techniques that teachers could use for the effective learning of biology by the students.

Teaching and Learning Activities

Teaching and learning activities can be grouped into two types:

- Those where the learner is actively involved in the learning process. The student has actually to physically and mentally do something in order to increase his or her knowledge (active Learning); and
- Those where the student 'absorbs' information by listening, reading or watching (passive learning).

Learning is more effective if students are actively involved in obtaining the information. Table 15 is a list of some teaching and learning activities used in biology lessons. Categorise them as either active (A) or passive (P) if properly used by the teacher: Write A or B in the column on the right in Table 10.1.

Table 10.1: Categories of Teaching and Learning Activities

Activity	Passive(P) or Active
1. Listening to the teacher	
2. Answering oral question	
3. Working together in groups	
4. Roleplay	
5. Class discussion	
6. Class debate	
7. Watching a science video	
8. Working on a computer program	
9. Watching a demonstration	
10. Taking down notes (from dictation, copying from the CB, copying from a worksheet	
11. Drawing conclusions from experimental results	
12. Looking up results	
13. Peer group teaching	
14. A circus of practical activities	
15. Conducting a project	
16. Writing own notes	

17. Preparing a poster for display	
18. Reading about biology	
19. Planning how to carry out an experiment	
20. Answering written questions	
21. Carrying out experiments	

Meaning of Teaching Method, Technique and Strategy

The choice of learning activities is determined by the *teaching approaches* adopted by the teacher. The approaches are categorised as teaching methods, teaching techniques and teaching strategies. These terms mean differently although they are all part of planning and teaching.

A **method** of teaching is a general approach to conducting a lesson, which involves a choice between whether learners will mainly be told or they will largely find out for themselves. There are two general methods:

- **Expository or Transmission** method; and
- **Problem-solving or discovery or inquiry** method.

The **expository** method is the approach where the teacher largely provides the information and the learner is the passive recipient. However, the problem-solving method is the approach where the teacher leads the learners to search for answers and information. In this process, the learners get the opportunity to develop requisite knowledge, skills and attitudes.

A teaching **technique** is a set of specific actions and processes through which the teaching method is realised (an activity performed to achieve the method). A given teaching technique may either be expository or problem-solving in nature. On the other hand, a teaching **strategy** is the sequencing or ordering of the techniques a teacher selects to teach the lesson.

Generally, a variety of teaching techniques increases the chances of reaching more students. At least two or more different techniques need to be used in a lesson to increase learners' participation. Table 10.2 indicates the relationship among the three approaches.

Table 10.2: A Scheme for Teaching Approaches in Biology

Method	Key Characteristics	Technique	Strategy
1. Expository	-Teacher oriented -Mainly passive -Mainly non-interactive -Content/ knowledge-based	-Lecture -Recitation -Discussion -Note taking -Workbook exercise -Teacher demonstration	Lecture then discussion then workbook exercise then recitation

		-Mainly Talk and Chalk -Promote surface learning	-Historical -Drilling	
2. Problem-solving		-Learner-oriented -Mainly active -Mainly interactive -Process-based -Mainly heuristic or discovery -Promote deep learning	-Class experiments -Debate -Panel discussion -Field course excursion -Written reports -Project -Games and simulations -Case study -Action research	Field experience then debate then projects then written reports

Factors Influencing Choice of Teaching Strategies in Biology

While selecting a method or technique or strategy of teaching biology, the teacher should assess its advantages and disadvantages in relation to others and also correlate them with the objectives of the lesson and the specific conditions in which learning is to take place. The factors which influence the choice of teaching methods, techniques and strategies include:

The teacher's mastery of content, experience, personality and communication skills

The effectiveness of teaching depends on the teacher's philosophy, style, attitudes and values, experience, and personality on one hand and grounding in pedagogy, familiarity with the subject matter, particularly the building blocks of the knowledge of science (such as facts, concepts, theories, laws and generalisations) as well as the scientific skills on the other hand. The teacher's choice of teaching techniques should be predicated on the value of such techniques in providing opportunities to the students to learn and practice valuable knowledge, skills and attitudes.

The size of the class

Enrolment in biology classes has been on the rise. The larger the class the less the teacher's use of teaching techniques that demand effective preparation and use of teaching resources such as class experiments. The teacher will tend to use techniques such as lecture and demonstration.

The student's entry behaviour – i.e. their intellectual or cognitive level, cognitive styles (information processing), age and practical experience, social and cultural backgrounds. Until and unless students have acquired such skills they will not learn using some techniques such as practical work in the laboratory.

Nature of content to be taught

The cognitive demand of the subject matter influences the choice of teaching techniques. The difficulty level of the subject matter depends on the type of the building blocks of its knowledge (such as facts, concepts, theories, laws and generalisations) as well as the scientific skills it purposes to promote. For example, if the concepts are mainly theoretical, many students will find it difficult to learn the subject.

Instructional Objectives

Instructional objectives determine the choice of teaching techniques. The ability of the teaching technique to facilitate the development of knowledge, skills and attitudes is particularly important in this choice. The instructional objectives to a large extent dictate the choice of instructional approaches to be used for each lesson. Many of the instructional objectives used in teaching biology address Bloom's cognitive levels, namely, knowledge, comprehension, application, analysis, synthesis, and evaluation. The first three address lower order thinking skills (LOTS) while the last three concern higher order thinking skills (HOTS). The higher order thinking skills require the students to select and use the most appropriate ideas, concepts, techniques, and procedures to address a new but unrelated situation and then modify or create a new hypothesis, idea, and tools, among others to help cope with the new situation. Teaching approaches closely relate to the LOTS and HOTS. The LOTS are promoted by the less interactive teaching techniques such as demonstration, discussion, lecture, and note taking among others. The HOTS are promoted by the more interactive approaches such as problem-based learning, project work, simulation and class experiments,

The objectives addressing the affective domain usually focus on the development of attitudes and values. These are preferably enhanced through the affective instructional techniques such as simulations, values clarification, debates, panel discussions, project work among others.

Availability of time and material resources

The resources may include equipment, apparatus, chemicals, commercial charts and other materials that may be expensive to procure. The availability of such resources is likely to be a limiting factor in the choice of teaching techniques. For example, the use of class experiments requires an abundant supply of such resources. The teacher may opt to use a demonstration to overcome the problem.

Transmission Method Techniques

Discussion

A discussion is an activity in which there is verbal interaction between teacher and students. It involves a free-flowing conversation, giving students an opportunity to express their opinions and ideas, and to hear those of their peers (Lemlech, 2010). There may be situations where the teacher does most of the talking, for instance when giving directions for doing an assignment or performing an experiment, or when introducing a lesson. A discussion may be a recitation where the teacher asks questions and solicits answers from the students or an exchange of ideas and questions between students or between students and teacher.

When is a discussion best used?

- Where students' opinions and experiences need to be known by the teacher, or are valuable and interesting to other students;
- Where the topic involves values, attitudes, feelings and awareness, rather than exclusively factual material; and
- Where it is necessary to give students practice in forming and evaluating opinions.

The discussion takes different forms; it may be a group or whole class discussion.

Whole Class Discussion

Whole class discussion is commonly used when the teacher addresses the whole class:

- At the start of the lesson to set the scene (i.e. introduction);
- During the progress of the lesson to help pupils develop concepts and/or skills (lesson development); and

- At the end of the lesson to draw together the key points (i.e. conclusion).

A good introduction to the lesson helps to get the pupils organised for the task at hand. They should know what they have to do and be keen to do the tasks. The exposition should involve questioning the pupils and allowing time for their contribution. Questions asked should cater for mixed ability so that all pupils in a class are actively involved at this stage to capture their interest in the lesson.

During lesson development, the teacher should involve pupils adequately. This is best done through several activities such as the use of worksheets that require them to fill in blacks in a page of notes. The teacher should, therefore, prepare well for what pupils will be doing during the lesson. However, time spent in whole class discussion during this stage should be limited, giving time for individual students to work on their own.

Group Work Discussion

Group work is best suited in the following circumstances:

- When pupils are expected to be involved in **practical tasks** in the laboratory, where groups of two or more work together, group work is essential. In this case, students work either on identical practical work or different aspects of the same topic. This seems to be largely used as a variant of whole-class teaching, as little use is made of a differentiated approach, or of the different learning experiences which group work can give. Indeed, many biology teachers would prefer individual practical work if the class were small enough and time and apparatus allowed it. Group work is not, in such a case, being used as a specific teaching strategy to achieve particular objectives, but as a way of organising a class and equipment available for the class activity;
- When students are working on **a project**, quite often they do so in groups;
- A group discussion of **a problem** or of **questions** during a lesson/topic, for example, finding out as to why a particular phenomenon occurs, is best addressed in group work. This also helps find out what pupils know;
- When students are expected to conduct a group preparation of work and materials to be presented later;

- Group work is handy at the start of a topic through which the teacher ascertains students' pre-conceived notions of the subject matter or topic;
- Group work is also commonly used at the end of lesson/topic by presenting pupils with a new situation and asking them to explain it in terms of what they just learned;
- Group work is relevant when students are expected to be involved in planning for an investigation; each pupil comes up with ideas, then as a group agrees to a common procedure to be used; and
- During topic or lesson to determine the level of understanding.

Group Investigation or Practical Work

In the case of group investigation, the teacher's role is to facilitate inquiry and problem solving by motivating students, setting the stage for group work, and providing necessary resources. The teacher also monitors group work, helping them appropriately and evaluating their progress. The focus is on inquiry, problem-solving, sharing diverse viewpoints, and using democratic processes to make decisions. Students, although they need to be guided, they are allowed to exercise autonomy in structuring their inquiry.

Preparation before Group Work

For group work to be successful answers must be sought for the following questions:

- Why are you using groups in this lesson? Is it to encourage social skills or some other reasons?
- How big will the groups be and how will they be formed? Do you need to reorganise the furniture and teaching space to help?
- How do you intend to start the lesson-setting the scene, motivating the class, giving instructions, etc.?
- How will you ensure that each group has worked at the right level, and that chemistry content and activities are geared to the groups' abilities?
- What provision have you made for groups who finish early?
- What is your role-teacher, motivator, helper, and organiser?
- Have you planned for a good end to the lesson which will draw the class together-pooling and sharing of experiences, commenting on things which went well or badly, reporting back from the groups,

explaining noteworthy or spectacular outcomes, assessing what has been learned, looking forward to the next lesson, etc.?

Box 10.1

Discussion Questions

1. Identify the barriers to successful group work for a given biology lesson.
2. What are some of the measures you would take to make group work successful in a biology lesson?

Advantages of Group Work

The following are the key advantages that accrue from using group work in a biology lesson:

- Group work allows teachers to set differentiated tasks to pupils of different abilities. Each group might contain pupils of similar ability working on similar or identical tasks;
- It is an excellent approach to mixed-ability teaching and may help pupils progress at a rate determined by their ability. While a low-ability group might struggle to get through the task set for them, a high-ability group might race through the original set of work and then go on to do extension materials;
- Alternatively, pupils may be grouped in a mixed-ability arrangement so that the individuals can learn from one another by exchanging ideas and views and helping each other. This is the essence of constructive alignment;
- With careful planning, the teacher working with a group can stimulate higher levels of thinking and challenge students to predict, hypothesize, and draw conclusions;
- Encourages cooperation and leadership among students;
- Enhances the development of self-esteem and self-confidence of both slow and fast learners; and
- Helps students to develop skills such as effective communication, critical thinking, problem-solving, interpersonal relationship, and decision making.

Homework or Assignment

The status given to homework varies from school to school. In some schools, it is compulsory for all students to have homework diaries that are signed to ensure the work is done. In others, there may be little pressure to do homework.

Advantages of Using Homework

Some of the reasons for setting homework are that it:

- Provides pupils with the opportunity to work on their own, without interference from elsewhere, and produce something that they can value;
- Gives a chance to practice techniques and learn information at their own pace;
- Helps develop the skills necessary to become effective independent learners;
- Gives the opportunity to reflect on work done in class and identify areas of work not fully understood;
- Creates an opening for parents to become more involved with their children's work; and
- Creates the opportunity to do further work and make more rapid progress through the curriculum.

Improving the Effectiveness of Homework

For pupils to get maximum benefit from homework the pupils:

- Need clear guidelines about how to carry out the tasks and some indication as to the nature of the product they are required to produce;
- Should be given a rough approximation as to how long they should spend on their homework;
- Need to have a system/structure for delivering and collecting homework;
- Should be in a position to clearly relate it to the work they do in school. If pupils see that the work has a purpose then they are more likely to tackle it with enthusiasm;
- Should be given worthwhile tasks such as:

- Making personal notes from textbooks by reading the relevant potions;
- Answering questions on the worksheet using the information they already have in their exercise books;
- Writing up neatly an account where they only have rough notes of the work they have done in class;
- Writing a piece of continuous prose, such as their opinions on particular issues or writing a letter to a friend describing their work in chemistry;
- Formulating hypotheses and experimental designs from a given problem situation;
- Planning an investigation;
- Writing up the results and conclusions of laboratory and demonstration experiments;
- Carrying out a survey amongst family and friends, linked to class work;
- Finding out things about their home;
- Keeping a diary of their diet/exercise;
- Preparing a poster;
- Revising for a test; and
- Carrying out a DART activity – "Directed Activities Related to Text."

Box 10.2

Discussion Questions

1. How does your school encourage pupils to do homework and how does it equip them with the necessary study skills to carry out their tasks?
2. Is there a standard procedure for setting and marking homework in the biology department?
3. Suggest ways you may put in place to ensure pupils effectively benefit from biology homework.

Demonstration

In this technique, the teacher shows, displays or demonstrates experiments or skill as pupils observe and record information. Its purpose is to show or illustrate or explain or clarify a procedure, process or phenomenon quickly and economically.

Advantages of using Demonstrations

- Provides the learners with the opportunity to develop a spirit of curiosity, self-reliance and inquiring mind;
- Provides learners with opportunities to develop cognitive skills such as observation, raising questions, identification, recording, comparing and drawing conclusions; and
- Control of the pace of the learning process by the teacher.

Demonstrations are preferred when

- Time is inadequate for the task;
- Techniques to be used are rather complex;
- Equipment or material for the whole class is inadequate or too expensive or too delicate;
- Chemicals or apparatus to be used are dangerous for students to handle;
- It can be used to initiate activity for a lesson to help pupils raise questions or solve the problem – situations or to clinch the idea to be used in the lesson;
- There's a need to verify facts and principles or giving new information to students; and
- There's a need to develop specific scientific skills such as measuring, drawing a liquid using a straw, etc.

Making Demonstrations Effective

When demonstrations are used, the teacher should:

- Try the activities out in advance to ensure they work;

- Ensure there are spare materials in case of damage to the major ones;
- Ensure each student has a clear view of the demonstration. This is made possible if the students sit in a U-shape around the demonstration table (Figure 10.1);
- Ensure that pupils are fully involved in the ensuing discussions, in making observations, in recording important information and in raising and answering relevant questions;
- Repeating steps that fail; and
- Observing safety precautions.

Question and Answer

The best way to involve pupils in the learning process is through questions. It is, therefore, necessary that you should pay particular attention to the type of questions you ask. Always think about the way you expect learners to respond when you ask a question.

There are several different types of questioning which elicit different forms of response from pupils. In a typical lesson, you may use questions to create interest in the topic, find out what pupils already know, stimulate thinking or to find out how far they have understood the lesson.

Closed questioning is commonly used to get a simple, often anticipated response, for example, "what is the colour of the Hibiscus leaves?" By asking closed questions you can find out:

- If the pupils remember what they learned yesterday or during the lesson
- If the pupils have understood the lesson
- If the pupils know enough to be able to solve similar problems.

On the other hand, open questions are those where a pupil might offer any one of a range of answers or explanations. Such questions help pupils to develop thinking skills as they allow them to guess, give an opinion, a reason or an explanation from the information they already have. It is important that pupils are encouraged to express their own ideas through open questions.

Open-ended Questions include:

- What do you think happens next in the story?
- What will happen to our garden bean seedlings if there is no water?

- Why do the seeds fail to germinate in the absence of oxygen?
- How can we explain the appearance of a rainbow?

As you work with the pupils you should increasingly expect them to justify their responses, thus developing their thinking skills.

Using Questions Effectively

As you prepare your lesson plan give some thought to the questions you want to ask and how you are going to ask them. You will get the best results if you think of a variety of key questions and prepare them as part of the lesson plan. Avoid the questions that elicit yes or no answers and use questions that encourage pupils to think and solve problems. The following hints are useful in asking questions:

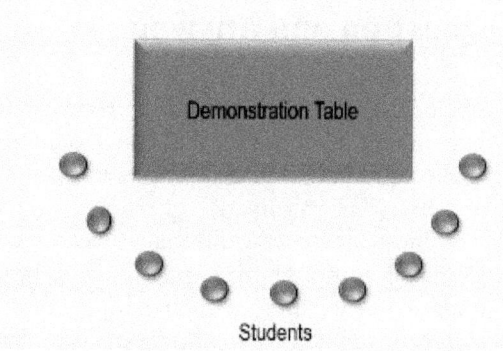

Class Set Up in a Demonstration

- Use a range of question types;
- When you want to increase the pace of an oral session use closed questions;
- When you want to explore children's thinking and to encourage their explanations, use more open questions;
- To promote consolidation of understanding ask children to talk about what they have done;
- If pupils are stuck, use a 'prompt' question, i.e. ask the question in another way';
- Give every pupil a chance to answer questions at some time during each day;
- Avoid chorus answers;
- Be ready to admit when you don't know the answer yourself; and
- Avoid using scorn and ridicule if the answer is incorrect; praise correct answers.

Advantages of using Questioning in Teaching Biology

The use of questions in a biology lesson provides the following advantages:

- Encourage students to develop thinking skills as questions allow them to guess, give an opinion, a reason, or an explanation;
- Encourage students to develop problem-solving skills;
- Encourage students to express themselves in terms of providing their own ideas;
- Promote consolidation of understanding by asking students to talk about what they have done;
- Creating interest in the topic;
- Finding out what pupils already know or to find out how far they have understood the lesson;
- Stimulating thinking; and
- Involving pupils in the learning process or improving their rate of participation in the learning process.

Lecture

A lecture is a didactic instructional technique involving one-way communication from the active presenter to a more-or-less passive audience. It is less interactive and less effective in promoting effective learning of biology.

Advantages of Lecture Technique

- It provides an opportunity for the teacher to give a large volume of verbal information that is not available to students in other forms;
- Effective in transmitting factual knowledge and understanding of basic facts and concepts;
- It helps the teacher to cover a lot of material within a short period of time;
- The authority of the teacher is usually preserved; and
- Appropriate for achieving low-level cognitive objectives.

Disadvantages of Lecture Technique

- It is highly non-interactive;
- Students have a limited attention span;

- Has a low yield in terms of learning outcomes;
- The student is often a passive recipient in the learning process;
- It is highly teacher centred;
- It emphasises on the transmission of content at the expense of providing an opportunity for the development of skills and attitudes.

Improving the Effectiveness of Lecture Technique

- Talking clearly and at a reasonable pace while facing the students;
- Structuring the presentation to provide stimulus variation such as the use of resources (such as PowerPoint), and question/answer sessions;
- Keep lectures short interspersed with other activities such as note-taking, observing a chart, drawing, answering questions in exercise books;
- Use of language that is at the level of the students; and
- Using a structured sequence in delivery e.g. Teacher Example on Blackboard (TEB) then Student Example on the Blackboard (SEB) followed by Student Example in Exercise Book (SEE).

Problem Solving Method: Some selected Techniques

Problem-solving is anchored in the inquiry. The inquiry is an active process in which the learner must connect what is new to past experiences and knowledge. The learner gets engaged in the discovery of concepts, testing ideas, and organising and structuring new knowledge in a constructive manner (Lemlech, 2010). The inquiry has its roots in reflective thinking and critical thinking.

Reflective thinking is purposive thought that is directed and controlled. It involves looking for a solution to a problem and being critical of the evidence so as to arrive at a balanced solution.

Inquiry teaching involves helping students to select a problem, formulate questions, and suggest possible hypotheses. According to Dewey (Cited in Lemlech, 2010), reflective thinking is defined as "**Active, persistent, and careful consideration of any belief or supposed form of knowledge in light of the grounds that support it and the further conclusions to which it tends**"

Inquiry teaching is important for several reasons:

- Students are motivated to search for answers to the questions they raise, especially by searching for relatedness, regularities and patterns. They become engaged in the process of reflective thinking;
- Students may often discover new ideas in the inquiry process;
- Students develop problem-solving skills; and
- Students learn to organise their thinking process so that they connect their ideas and manage their time in a better manner.

The inquiry-oriented classroom is characterised by the following features:

- Teacher and students are able to sustain reflection/reflective thinking;
- Questions by teacher stimulate discussion;
- Questions are structured to guide students' search by raising their level of thought; and
- Providing materials that direct students' hypotheses formulation and thinking.

The role of the teacher in the inquiry-oriented classroom includes the following:

- Facilitator of students' learning;
- Arouses interest and helps students raise questions so that they inquire and learn to draw their own experiences;
- Helping student to perceive discrepancies in their information; and
- Sets classroom environment in which students gain confidence in directing own thinking and investigation.

Creating an inquiry-oriented environment entails the following:

- Providing a physical and social environment that encourages students to develop inquiry skills, especially arrangement of the classroom, planning strategies, and selection of materials; and
- Allowing students to interact with each other in groups and with the materials for the purpose of engaging in discovery learning.

The teaching techniques that follow, if properly planned for and executed, will promote students' engagement in the inquiry.

Practical Work or Class Experiments in Biology

Practical work is an essential component of the teaching and learning of biology. Students can be stuffed with facts and theories but without experiments, they cannot experience the reality of biology as a science. However, such an approach to teaching biology involves considerable time, effort and expense.

Advantages of Practical Work in Biology

This approach helps to:

- Develop in the students, useful scientific skills such as accurate observation, careful recording, measurement, interpretation of data, as well as manipulative skills;
- Create a permanent impression of knowledge in their minds, of what is learned through experimentation;
- Learn by doing where they are active participants (pupil-centred approach);
- Make it possible for students to understand complex, abstract ideas and thus develop chemistry concepts and principles;
- Gives students an opportunity to participate in and have an appreciation for the scientific method;
- Train students in problem-solving and thus promote the development of logical thinking skills;
- Give insight into the scientific method and develop expertise in using it.
- Make scientific phenomena more real/enhance understanding of scientific ideas;
- It sheds light on theory (the idea of seeing is believing);
- Arouse and maintain interest in chemistry learning;
- Train in laboratory skills; and
- Develop useful scientific attitudes such as curiosity, tolerance, patience, open-mindedness, objectivity, cooperation, and willingness to suspend judgment.

When is Practical Work used

- When students are expected to learn basic practical skills, e.g. using the Bunsen burner, how to read measuring instruments accurately, how to use a microscope, etc.;
- When illustrating a theory or concept – i.e. understanding a scientific idea better by observing an experiment illustrating that idea;
- When proving a theory – as when pupils go through tasks on a worksheet or in a workbook intended to prove a given theory. Critics of this approach argue that this reduces biology to little more than 'cookery', with pupils ardently following a recipe; and
- When investigating ideas practically. Investigations are more mentally taxing, involve more than just carrying out an experiment and require pupils to plan their own experiments, execute them, evaluate them, look up for information, and make conclusions.

Making Practical Work in Biology Effective

Practical work in biology becomes more effective for learning by ensuring that:

- The objective of each practical is very specific. What data is to be collected, how the results will be presented, the relevant deductions and conclusions, materials and apparatus, and procedures to be followed;
- Conducting the practical in groups of between 3-5 students in order to: make it cost effective in terms of materials, time and equipment; promote sharing of ideas; promote teamwork and team spirit; stimulating learning, and develop communication skills;
- Providing clear instructions;
- Demonstrating the crucial steps in the procedure for the learners to pay attention to safety, precision, and care of the equipment;
- Trying out the experiments in advance to ensure they work within the stipulated time and to identify anticipated difficulties and challenges;
- Ensure the requirements are appropriately labelled;
- Supervise and guide the students appropriately during the practical; and
- Encourage students to evaluate their practical work in terms of the accuracy of the results and suggestions for improvement.

Planning for class experiments in Biology

Practical work is best undertaken in a systematic manner. The sequence of activities in an investigation is as follows:

a) **Setting the scene:** Through whole class discussion the teacher states the nature of the problem to be investigated and states the objectives of the planned experiment. The teacher breaks the class into groups for the purpose of the experiments.

b) **Planning Investigation:** Through group work, the students undertake activities guided by the following questions:
   - What do we understand about this work?
   - What do we want to find out?
   - What do we need in this investigation?
   - What strategies could we employ to solve our problems?
   - What are the variables? Which variables do we investigate?
   - Which variables do we need to control?

c) **Designing the experiment:** Through group work, the students undertake the following activities:

   - Clarifying the meaning of the variables, namely independent, dependent and control variables;
   - Deciding on a plan to carry out; and
   - Modifying the plan if necessary.

d) **Performance of the investigation:** The students conduct the experiment in their groups.
   - Setting up the apparatus;
   - Conducting the experiment;
   - Collecting data; and
   - Recording data.

e) **Discussion of results and evaluation of the experiment:** The students in their groups interpret the results and evaluate the success of the experiment. They do that by undertaking the following activities:

- Highlighting the key results;
- Writing an explanation of the results using scientific ideas, if possible;
- Writing a comment on how the investigation could be improved; and
- Writing a conclusion that matches the results.

f) **Writing the practical report:** Each student writes the report of the experiment based on the guidelines provided by the teacher;

g) **Whole class discussion:** The teacher leads the students in a whole class discussion to summarise and consolidate what has been learned from this experience.

Factors to consider in Planning Practical Work in Biology

In this planning process a number of factors need to be put into consideration:

- The composition of the class in terms of the abilities of the individual pupils. Certain activities are more effective with groups of pupils of similar ability whereas others work well with mixed – ability classes;
- Size of the class: some practical work is only possible with small classes, while other activities work best for large classes;
- The quality and quantity of apparatus, equipment and materials may limit the extent to which students undertake the practical activities;
- The layout of the room may make it difficult to conduct certain activities such as group work;
- The level of students' motivation that ensures their sense of excitement and interest in the lesson; and
- The time available for the students to undertake practical activities.

Box 10.3

Discussion

You wish to introduce the concept of genetic variation in human beings. You will need to do this using members of your class of about 50 students. You will also need copies of the main groups of fingerprints and measuring tapes or rulers. You will need the students to investigate the following characteristics: Tongue rolling; Fingerprints; and Height.

a) Prepare instructions to guide the students in conducting the three investigations.
b) Indicate the type of data required and how the students will record the data.
c) Prepare a worksheet that guides the students to distinguish between continuous and discontinuous variation.

The 5E Instructional Model in Biology

The 5E Instructional Model was developed in the USA in 1989 by the "Biological Science Curriculum Study (BSCS)" group led by Roger Bybee. The model was designed to provide teachers with a teaching model that motivated inquiry thinking, group interaction, and active participation of the students in the learning process. The National Space Agency (NASA), later developed elaborate curricula materials to support the model in schools. The materials exist as NASA eClips™. A summary of activities in the model is given in Table 10.5.

The Model describes an interactive teaching sequence that can be used in a lesson in which the teacher acts as a provocative stimulator for inquiry thinking and as a consultant to help students refocus their thinking. The 5Es represent five stages of a sequence for teaching and learning, namely: Engage, Explore, Explain, Extend (or Elaborate), and Evaluate. Each of these is explained as follows:

Engage: This stage aims at capturing students' interest and get them individually involved in the lesson. The students' prior knowledge is diagnosed and their interest engaged in the concept or phenomenon for the lesson. During this experience, students first encounter and identify the instructional task. The task is designed to arouse students' thinking that encourages them to begin asking their own questions. Students make connections between past and present learning experiences; they set up the organisational groundwork for upcoming activities. In the BSCS, NASA eClips™ videos were used designed to engage students through observation and discussions.

For example, the students are introduced to the various anomalies in human beings, including albinism, sickle-cell anaemia, haemophilia, and colour blindness. The use of pictures, photographs and video clips could make a lasting impression and appreciation of these genetic problems. The question to be explored is: What causes these genetic anomalies in human beings?

Explore: In this stage, the students participate in an inquiry-based activity that facilitates conceptual change. The students manipulate and interact with materials designed to provide them with a chance to build their own understanding. The inquiry process focuses on questioning, planning the investigation, setting up apparatus to conduct an investigation, data collection and analysis, critical thinking, drawing inferences and conclusions. As they work together in groups, students build a set of common experiences which prompt sharing and communicating. The teacher acts as a facilitator, providing materials and guiding the students' focus. In the BSCS, the NASA eClips™ was designed to help students **explore** new topics on their own; they encouraged students to make hypotheses, test their hypotheses, draw inferences, make conclusions, and communicate the findings.

For example, analogies of gene mutation are used to illustrate how genes provide inaccurate coded messages that lead to genetic problems. The students use a table such as Table 10.4 to indicate simple messages that illustrate the changes that cause distortions in the intended messages. They are asked to practice combinations of messages such as those shown in the table. They are also asked to relate the illustration to what happens to genes or strands of DNA thus causing various types of gene mutations. The students may now suggest other forms of mutations, namely, chromosomal mutations, and induced mutations such as polyploidy that are used in enhancing agricultural production.

Table 10.4: An Illustration of Genetic Mutation

Intended message	Actual message	Type of mutation
Give me a spoon	Give be a spoon	Substitution
Teacher is not coming	Teacher is coming	Deletion
This is my mate	This is my team	Inversion
John is praying	John is spraying	Insertion

Explain: At this stage, the students generate an explanation of the phenomenon they are learning by providing them with an opportunity to communicate what they have learned so far and figure out what it means. Communication occurs between peers, with the facilitator, and through the reflective process. In the BSCS programme, once students build their own understanding, they may use NASA eClips™ to help summarise or explain their own ideas and to correct their misconceptions.

For example, students may look at induced mutations more critically by suggesting ways of reducing frequent exposure to materials that may cause mutations such as exposure to x-rays and radioactive materials during medical diagnosis and handling of e-waste.

Extend/Elaborate: At this stage, the students' understanding of the phenomenon is challenged and deepened through new experiences so that they are allowed to expand on the concepts they have learned, make connections to other related concepts, and use their new knowledge in relevant applications.

For example, students may elaborate more on the phenomenon of polyploidy in plants as a chromosomal rather than mutation of the genes, which is usually induced for advantageous reasons in agriculture.

Evaluate: During this final stage, student assess their understanding of the concepts and ideas. The purpose of the evaluation stage is for both students and teachers to determine how much learning and understanding has taken place. However, evaluation and assessment can occur at all points along the continuum of the instructional process. Some of the tools that assist in this diagnostic process are drawings, models, teacher observation, student interviews, portfolios, journals, projects, performance tasks and problem-based learning products.

For example, students are asked to illustrate gene mutations using diagrams to illustrate their degree of understanding of the concept. In this case, they are expected to make the illustrations in terms of the changes in the genetic code of DNA.

Table 10.5: Summary of Activities in the 5E Instructional Model

STEP	TEACHER	STUDENTS	ANALYSIS
Engagement	-Captures students" interest and attention using suitable media such as pictures, video -Raises questions to elicit students' thinking -Leads students into groups to solve a problem	-Students respond to the teacher's questions and raise further questions	-Students are motivated through the use of pictures, etc. -Choice of media creates interest and elicits students' sensitivity and concern
Exploration	-Directs the discussions in groups; each group asked to report to the whole class-	-Students hold discussions in groups and report their findings	-Students ask further questions focusing on solutions in an interactive manner.

	-Consensus leads to some conclusion		-This phase is similar to group work phase of group investigation
Explanation	-Leads students to individual seats to provide answers or explanations to what they observed in groups	-Students provide their group reports	-Provide further insight using past experience and knowledge
Elaboration	-Lead students to further elaborate on the students' responses and observations	-Students provide further explanations to responses and observations	-Integration of what is learned with what is known -Students expand their thinking
Evaluation	-Lead students to put their discussion, group work, and readings together—like talking to parents about contribution to problem solution	-Students consolidate their findings together and sort out gaps	-Drawing conclusions -Application of knowledge -Planning applications

Relationship of 5E Instructional Model to Constructivism

5E Instructional Model is anchored in the Constructivist view of inquiry-based learning. The model is based on the idea of students asking questions, seeking out solutions to problems, and using their previous knowledge to construct new knowledge. It is thus rooted in Constructivism and is intended to help students build their own understanding from experiences and new ideas.

Since the late 1980s, BSCS has used the 5E Instructional Model extensively in the development of new curriculum materials and professional development experiences in biology. The Model also enjoys widespread use beyond BSCS for curriculum frameworks, assessment guidelines, course outlines, curriculum materials, and teacher professional development training materials. Other models have been adapted from this model including the 6E and 7E models.

Advantages of Using 5E Instructional Model

- Develops in the students scientific skills such as accurate observation, careful recording, measurement, interpretation of data, as well as manipulative skills;
- Creates a permanent impression of knowledge in the students' minds, of what is learned through experimentation;
- Promotes learning by doing where students are active participants (pupil-centred approach);
- Makes it possible for students to understand complex, abstract ideas and thus develop biological concepts and principles;
- Encourages development of useful scientific attitudes such as raising questions, curiosity, withholding judgment until the evidence is adduced, among others;
- Trains students in problem-solving and thus promote logical thinking skills;
- Make scientific phenomena more real;
- Enhances students' understanding of scientific ideas. It sheds light on theory (the idea of seeing is believing);
- Arouses and maintains students' interest in biology learning; and
- Develops in the students useful scientific attitudes such as curiosity, open-mindedness, objectivity, cooperation, and willingness to suspend judgment.

Problem Based Learning in Biology

Problem-based learning (PBL) is a student-centred approach in which students actively learn about an idea or a concept through the experience of problem-solving. Students learn both thinking strategies and relevant knowledge. Problem-based learning differs from traditional approaches. The process begins by introducing the students in a group to a problem then encouraging them to use prior knowledge to develop new ideas and to discuss thoughts. These processes lead to a whole spectrum of ideas about how to solve the problem and the actions that will be taken to reach the solution.

PBL was first used in training medical personnel at McMaster University in Canada in the late 1960s by Howard Barrows and his colleagues. It spread to the other medical schools in various parts of the world and in 2008; it was used by Parramatta Marist High School in

Australia. It has since been part of teaching and learning in schools around the world.

Barrows (1996) identifies six basic features of Problem-Based Learning. These are as follows:

- PBL is student centred;
- Active learning or interactive learning process;
- Learning is done in small groups of 6-10 students;
- The teacher is a facilitator or guider of the learning process;
- A problem forms the basis for the organised focus of the group, and stimulates learning;
- The problem is a vehicle for the development of problem-solving skills. It stimulates the cognitive process; and
- New knowledge is obtained through Self-Directed Learning (SDL).

Problem Based Learning Process

Students work in groups to identify what they already know, what they need to know, and how and where to access new information to be able to solve the problem at hand. The role of the teacher is that of facilitating learning by challenging, supporting, guiding, and monitoring the learning process. Feedback and reflection on the learning process and group dynamics are essential components of PBL. The teacher builds students' confidence to be able to take the plunge into the PBL agendum.

The problem-based learning process is achieved through the following steps:

- Learners are presented with a problem and through group discussion, activate their prior knowledge;
- Through group work, students develop possible theories or hypotheses to explain the problem. Together they identify learning issues to be researched. They construct a shared primary model to explain the problem at hand;
- The teacher as a facilitator provides a scaffold, which is a framework on which students can construct knowledge relating to the problem;
- After the initial teamwork, students work independently in the self-directed study to seek answers to the issues and questions identified; and
- The students re-group to discuss their findings and refine their initial explanations based on what they learned.

Learning Procedures for Problem Based Learning

- Read and analyse the problem scenario in line with the learning issues;
- List hypotheses;
- List ideas, hunches;
- List what is already known;
- List what is unknown;
- List what needs to be done; and
- Develop a statement problem and the problem format.

Advantages of using Problem Based Learning

- Encourages students to develop effective problem-solving skills;
- Development of scientific skills;
- Promotes self-directed learning;
- Stimulates critical reasoning;
- Stimulates elaboration;
- Enables students to be involved in self-directed learning;
- Development of effective collaborative skills;
- Enhances interest in the learning process;
- Development of useful scientific attitudes such as cooperation;
- Improves students' memory and lessens instances of forgetfulness;
- Promotes positive attributes of communication, teamwork, respect and collaboration. These skills provide for better future skills preparation in the ever-changing information explosion world;
- The process is designed to work best in a student-centred and group-based approach; and
- Allows for the integration of innovative pedagogical approaches and the use of technology tools to promote learning. Examples of these techniques include strategic use of questioning to facilitate discussion, providing prompts for reflection, and requiring peer review as part of the reporting and assessment processes.

Disadvantages of using Problem Based Learning

- Active problem-solving in the form of PBL may be difficult for slow learners if it's without proper use of scaffolds on the part of the teacher (Sweller and Cooper, 1985; Cooper and Sweller, 1987)

due to a high cognitive demand and high cognitive load imposed by PBL;
- Many learners may find it difficult to process a large amount of information required for problem-solving in PBL in a short amount of time. Once learners gain expertise the scaffolding inherent in problem-based learning helps learners avoid these difficulties;
- Implementation of PBL learning in comparison to traditional instructional learning is a challenge on the part of the teachers. Various factors can influence the implementation of PBL: the extent of PBL incorporation into the curriculum, group dynamics, nature of problems used, facilitator influence on the group, and the motivation of the learners;
- Provision and Management of learning resources and facilities that support self-directed learning may prove to be a big challenge for many schools;
- Infusing PBL in the mainstream curriculum requires adequate knowledge building, written and interpersonal interactions and the experience of the problem-solving process on the part of the teachers. This may not be readily achieved by many teachers;
- When students explore complex problems in a PBL environment, complex problems may generate a heavy load on students' working memory because of their lack of proper schemas to integrate new information with their prior knowledge (Kirschner et al., 2006). Given the complex nature of PBL, it is critical that instructors provide guidance to learners through each of the PBL activities while providing "direct instruction on a just-in-time basis" (Hmelo-Silver, 2004); and
- Teachers feel frustrated with the difficulties of managing classroom implementation of PBL in an environment that requires daily and even hourly curriculum mapping of standards addressed and learning objectives assessed.

Relationship of Problem Based Learning to Constructivism

The common traditional teaching and learning of biology are often lecture-based. PBL represents a paradigm shift from this traditional teaching and learning philosophy that believes that learners absorb knowledge (Schmidt, et al, 2011). Its constructs for teaching and learning are very different from traditional classroom teaching.

PBL can be considered a constructivist approach to instruction for the following basic reasons:

- It emphasises collaborative and self-directed inquiry-based learning supported by flexible teacher scaffolding (Schmidt, et al, 2007) that allows students to construct knowledge;
- The teacher does not lecture to students; rather it is based on the idea of students asking questions, seeking out solutions to problems, and using their previous knowledge to construct new knowledge;
- Students are active agents who engage in social knowledge construction;
- Through questioning, students are expected to identify a problem and figure out their own ways of solving it based on their previous knowledge;
- Students take more responsibility for their own learning. Teachers become facilitators or guides, rather than lecturers who do not interact with the students. Collaboration and group work are emphasised;
- Not all answers are definite and absolute; there is room for discussion and problem-solving. Tasks and activities have relevance to the learner;
- PBL assists in processes of creating meaning and building personal interpretations of the world based on experiences and interaction; and
- PBL assists to guide the student from theory to practice during their journey through solving the problem.

Example of Problem Based Learning Activity

The Problem

Wanjiru, in an effort to understand more about new ways of teaching biology and to be able to teach more effectively, approached the Church Official in charge of Education services for financial support. In the affirmative, Wanjiru was asked to look for a suitable short course or seminar or workshop to attend. Fortunately for Wanjiru, she had already come across an advertisement in the *Daily Nation*, which she showed to the official.

The advertisement carried some of the following details:

XYZ Staff Development Solutions

We are a reputable staff development consultant group. We shall be holding a staff development training seminar between 1st April to 8th April, 2010. The details are as follows:

Contact:

School-based curriculum development; setting clear targets for teaching; setting clear targets for professional development; effective class management; Drawing departmental improvement plan; identifying pupil learning challenges; Effective teaching methods and resources in biology; ICT in biology education; meeting targets for public accountability.

Target Group:

Secondary school personnel; Quality assurance officers, Curriculum support officers; refugee camps, arid and semi-arid schools, and internationally sponsored schools.

Duration: 7 days

Methods: *Group work, Multi-media demonstrations, plenary sessions; Field course excursions and practical work demonstrations.*

After carefully scrutinising the advertisement, the official requested Wanjiru to immediately write to the Church requesting for sponsorship to the seminar. However, Wanjiru was strongly advised that her application stood a better chance for consideration if she provided a justification for the attendance of the seminar.

Box 10.4

Discussion

Your opportunities for funding by the employer are most likely to be met where your needs are closely aligned with those of the institution-wide improvement. Assuming that your school wishes you to develop pedagogical skills, provide a justification for your attendance of the seminar. Your justification should be based on your understanding of biology teaching so far and the extent to which the seminar will help you fill the gaps if any, in this understanding.

Construct your proposal indicating your justification

Project Based Learning in Biology

Project work is an extended piece of practical work, usually conducted independently or in groups, in the school environment or away from the school environment. It is different from class experiments since it involves students undertaking different manipulated activities or varied library search or different problem-solving tasks independently either as individuals or in groups. In laboratory work, students usually undertake the same practical tasks in the same place at the same time under the direct supervision of the teacher.

Advantages of Project Based Learning

Project work augments students' learning and has the following advantages:

- It gives the students an opportunity to relate biological knowledge (concepts, ideas, principles, generalisations and theories) learned in the class or laboratory to the real life in the environment;
- It gives the students the opportunity to work with experts in the field and with members of the community surrounding the school;
- It cultivates students' interest in biology;
- It enables students to develop and use a wide range of scientific skills (both cognitive and psychomotor) necessary for solving problems encountered in their everyday lives;
- It trains students in the art of the scientific method in terms of raising questions, hypothesising, experimenting, and making conclusions, among other skills;
- It helps students to develop communication skills in terms of recording information and writing reports; and
- It promotes an interdisciplinary approach to learning biology as it calls for use of mathematics skills, technology, and economics among other areas of concern.

When Project Work should be used

Project work is useful in the following situations:

- When the topic requires observation and monitoring of certain aspects over a long period of time, for example, growth and devel-

opment of plants and animals as in the life cycle of insects, and germination, growth and development of seedlings;
- When we need an understanding of life mechanisms such as locomotion in fish and arthropods;
- When we want to study the factors affecting certain biological processes such as photosynthesis, respiration, osmosis among others;
- Attempting to relate soil types to growth and development of plants;
- When we need to understand the sources and effects of pollution of the nearby river or stream by detergents, fertilizers and siltation; and
- When there is a need to understand the availability and uses of biological resources in the community.

Designing Projects

The project may be experimental, observational or survey design (Twoli, 2006). Whichever design is selected, the project must revolve around five main criteria, namely: centrality; driving question; constructive investigations; autonomy; and relevancy.

First, centrality defines everything the student needs to learn and should be centred on the project that the student is working on.

Second, the teacher must ensure that a driving question is identified which the student must answer through completing the project. This question is the most important aspect to a successful project-based learning process because it is what motivates the student to learn as well as providing the student with an idea of what knowledge is expected to be learned from the project. Also, the question cannot have a predetermined outcome, it should require the student to use prior knowledge to come up with a conclusion.

Third, constructive investigations are arrived at if the students plan investigation based on the ideals of experimentation, involving identification of variables and control of some of them.

Fourth, the students should be given the independence or autonomy they need to conduct the project to its logical conclusion.

Fifth, the project should relate to the students' environment (relevancy) as the basis upon which to construct new knowledge. All of these criteria are important for the students because they help to engage them in the project as well as to allow them to make connections to real-world situations.

Making Projects Effective

- Students must plan carefully before embarking on the project, for example identifying the objectives, central concepts and the research question to be answered;
- Ensure that all the apparatus and other materials needed for the project are available before the beginning of the project;
- All the observations and measurements should be recorded appropriately in a file or notebook but not on separate pieces of paper that may disappear; and
- Attempt to pilot the study if time allows ensuring that the procedures and techniques to be used are appropriate.

Project Report

The report should be written carefully according to the guidelines provided by the teacher. The report is likely to take the format below:

- Title;
- Acknowledgement;
- List of contents;
- Abstract or summary of the project;
- Introduction: Background information, purpose, objectives, scope, and terminology used;
- Method and materials: Materials used, the design of the project, data collection, safety measures;
- Results: Either qualitative or quantitative and presented in suitable formats such as tables, graphs, drawings, and pictures;
- Discussion: Interpretation of results and implications to the people;
- Conclusions: What the results show and what can be derived from them;
- References; and
- Appendices: Instruments, chemicals and reagents used.

Sample Assessment Scheme for Project work in Biology

Assessment is necessary for the purpose of providing feedback to the students. A sample is shown in Table 10.6. You can always develop a suitable one for your use.

Table 10.6: Assessment Scheme for Students' Projects in Biology

Section of project	Marking points	Score
Abstract	Brief, clear giving: • Theme • Materials used • Key results and conclusions	0.5
Introduction	• Purpose/objectives of the study • the theme • Literature review • scope	20
Method	• Design • Data collection • Data recording • Safety measures	20
Results	• The accuracy of data/scales • Presentation of data • Analysis of data	20
Discussion	• Comparison of results with others • Implications for practice • Suggestions for further research	20
Conclusions	• Results relation with objectives • Theoretical explanation of results	10
Understanding of concepts and records	• References • Use in the project report	10
General organisation	• Formatting and organization of tables, graphs • Tidiness	0.5
TOTAL		100

Box 10.5

Discussion

Consider a biology project that you have recently carried out with your class and analyse the task for:

a) Procedural demand
b) Conceptual demand
c) The context in which it was set

How did you put into consideration the following factors to ensure the success of the project?

- The composition of the group. Certain activities are more effective with groups of pupils of similar ability whereas others work well with mixed – ability classes.
- Size of the group – some practical work is only possible with small classes, while other activities work best for large classes.
- Materials and apparatus

Motivation – ensuring pupils have a sense of excitement and interest lesson

Cooperative Learning

Cooperative learning builds on the advantages of a group discussion during the learning process. However, in cooperative learning, elaborate and challenging tasks that are well structured are provided to groups to engage them into active learning. The group is usually small and well structured. The students engage in hands-on activities that require a little more time for processing, analysis, reflection, and interpretation of experiences. In addition, the groups stay together for some considerable length of time and they acquire the status of social groups. The groups begin to collaborate to make sense from their experiences. Conversation, presentations, writing and concept mapping prompt the student groups to make their implicit ideas explicit and clarify their understanding. One fundamental outcome of this interaction is the creation of a **learning community**.

The **learning community** is as an organisation in which all members acquire new ideas and accept responsibility for developing and maintaining the organisation (Dianna, 2000). Every individual's contribution is significant to the life and well-being of the organisation (Argyris & Schon, 1996). The members share experiences, respect and trust each other, and enjoy an open environment for collaborative decision making. The members also feel that their insights are valued and taken into account in community life. Garvin (in De Vito, 1996) identifies the following five main activities of the learning organisation: Systemic problem solving; experimentation with new approaches; learning from experiences; learning from best practices of others, and transferring knowledge across the organisation quickly. Learning thus changes from individual competition to group collaboration of the twenty-first century.

When is Cooperative Learning used?

- Useful for teaching skills or specific information;
- Group members are responsible for helping each other to learn a specific skill; and
- The group is tested as a whole to monitor the groups' progress.

Cooperative learning and group investigation are similar in the following ways:

- Both rely on social interaction and group effort; and
- Both conclude with a class discussion that evaluates group processes and substantive progress.

Advantages of Cooperative Learning

The following advantages accrue from using cooperative learning in biology:

- Working in separate groups yields more ideas and potential solutions than do individual students;
- When students work in groups they act as teachers and peer teaching/learning becomes a powerful classroom process;
- The formal learning groups become social groups which collaborate to make sense from the various experiences;
- The well-structured and challenging tasks that are characteristic of cooperative learning lead to group engagement and active learning. The hands-on activities are accompanied by processing, analysis, reflection, and interpretation of experiences;
- The nature of the conversation, presentations, writing and concept mapping the learners engage in during cooperative learning prompts student groups to make their implicit ideas explicit and clarify their understanding; and
- Evaluation of cooperative learning is reliable since it is achieved through peer evaluations of each individual's contributions to a group effort based upon agreement among members.

Collaborative Learning

The teaching of biology is required to encourage collaboration. Collaboration refers to a holistic overview of the education process whereby individuals build on the value provided by others and enabling and empowering technologies. It includes students working with other students nationally, regionally or globally, incorporating team work using their team skills and interdisciplinary approach. Learners may also seek and work with experts as required. Students and teachers are collaborating across the world and beyond the time constraints of the teaching day.

Collaborative learning helps students by allowing them to focus more on real world problems as a basis of understanding. It also helps them to clarify understanding and support deeper learning. What they learn in one subject becomes useful in other subjects. Projects become more encompassing in an interdisciplinary manner. The sum of the students' learning becomes greater than the individual aspects taught in isolation. Teachers begin to network, collaborating with their peers in other subjects, to link and bind the learning in their subjects.

Conclusion

This chapter has emphasised that the teaching methods, techniques and strategies selected for use by the biology teacher determine the quality of learning by the students. To ensure quality learning by the students, the planning process must consider the best teaching and learning approaches and activities that enhance students' active and interactive engagement in the learning process so as to realise the instructional objectives. The learning activities which resonate the instructional approaches used should cater to the students' interests and needs and thus facilitate the learning process. The chapter has effectively outlined some of the best practice instructional techniques that teachers could use for the effective learning of biology by the students.

References

Argyris, C., & Schon, D. A. (1996). *Organizational Learning II*. Reading, MA: Addison Wesley.

Biggs, John (2003) *Teaching for Quality Learning at University* (2nd Edn). Berkshire: Open University Press

Diana B. Hiatt-Michael (2000). Schools as Learning Communities: A Vision for Organic School Reform. *Journal of School Reform*

Marton, F (1981) "Phenomenography – describing conceptions of the world around us."*Instructional Science*, 10: 177-200

Marton, F and Saljo, R (1976) "On qualitative differences in learning: I: out come and process", *British Journal of Educational Psychology*, 46: 4-11

Marton, F, Dall'Alba, G and Beaty, E (1993) Conceptions of learning, *International Journal of Educational Research*, 19: 277-300

Gagne, R.M and Driscoll, M.P (1988) *Essentials of Learning for Instruction* (2nd Edn) New Jersey: Prentice Hall.

A BIOLOGY LAB should be provided with standard equipment and apparatus

Photo: HKEOLA/FLICKR

11

CHAPTER ELEVEN

INSTRUCTIONAL RESOURCES IN BIOLOGY

◆ Definition of Instructional Resources ◆ Equipment and Chemicals for a Biology Laboratory ◆ SEPU Biology Kit ◆ Improvisation of Resources in Biology ◆ Preparation of Solutions in Biology ◆ Preservation of Specimens ◆ Live Specimens ◆ Biology Textbooks ◆ Teaching with ICT

Introduction

The psychological basis of using instructional resources during the learning process lies in the observation that they concretise the otherwise abstract concepts. The learning of biology is best achieved if the teacher uses the instructional resources to support the students' inquiry, as the student seeks 'the truth' about the problem that was driving the learning process. Exploration by students most effectively progresses when they have been well facilitated through the provision of learning resources equipped and guided during the learning process. This chapter focuses on the use of instructional resources ranging from simple materials to digital resources to facilitate the learning of biology.

Definition of Instructional Resources

Instructional resources are materials and facilities that provide information, expertise and support that make teaching and learning more meaningful and effective.

Types of Instructional Resources

The types of resources for teaching biology include:

a) Audio-visual Aids

Pictures, photographs, wall charts and posters, 3-D models, artifacts, newspaper cuttings of relevant specimens and related information, textbooks, reference books, mimeographs, micrographs, demonstration cards, radio tapes and cassettes for particular topics, videos for particular topics, films and film loops, spreadsheets and computer software, CD-ROM, worksheets, pamphlets from research stations and farmers training centres, transparencies, websites among others.

b) Realia

First-hand experiences such as live specimens, collections in herbaria, gardens, arboretums, museums, ponds, gardens, aquaria, among others.

c) Equipment, Apparatus and Chemicals

The biology laboratory should usually be provided with standard biology equipment and other support equipment, apparatus, and other materials. Support equipment includes micro-projectors, PowerPoint projectors, video players, refrigerators, radio cassette players, computers, while the standard biology laboratory equipment includes microscopes, egg incubators, fire extinguishers or sand buckets, hose pipe, cameras and autoclaves. The apparatus glassware, dissection kits, bio viewers, and hand lenses. The chemicals and reagents include indicators, iodine among others. The materials include bottle tops, tin cans and bottles, soil samples, anti-venom, fresh and preserved specimens, and staffed animal specimens.

Advantages of Using Resources to Teach Biology

- Provide real-life experiences for learners, thus making learning more meaningful and relevant;
- Concretise the otherwise abstract ideas, situations, and concepts so that learners understand them;
- Enhance the learning environment because learners are able to work in a physical as well as a mental medium;
- Provide a stimulating start to a lesson through presentation and closure phases as well as offering useful visual images that form permanent impressions in the minds of the learners. The resources, therefore, enhance remembrance of what is learned;
- Help learners to visualise processes and relationships that are normally invisible or difficult to understand (For example, spreadsheets create a graph that demonstrates a trend or shows whether one result is out of line with the rest);
- Provide an opportunity for students to reflect on events, situations, ideas in creating new mental models. This is because they see an actual event or visualise ideas and concepts and thus create their own analysis rather than reading someone else's description; and
- Make it possible for students and teachers to connect with other students, teachers, homes and communities regardless of physical distances.

Minimum Biology Equipment and Chemicals for a Biology Room or Laboratory

Where the school does not have a biology laboratory but a classroom converted to a science or biology room, the following list of equipment and chemicals would be fairly adequate (see Table 11.1).

Table 11.1: Minimum List of Equipment and Chemicals for Biology Laboratory

ITEM	DESCRIPTION	QUANTITY
1. Biology kit	Complete SEPU kit	10
2. Microscopes	Olympus HSC	5 or 10
3. Mortars and pestles	pair	10
4. Dissecting trays	Trays with wax	10
5. Blood lancets	packet	1
6. Dissecting kit	kit	10
7. Cork borer	Set of 12	1
8. Thermometer	Precision 110°C	10
9. Clinical thermometer	human	1
10. Petri dishes	Plastic/glass	40
11. Watch glass		10
12. Bunsen burners/spirit lamps/candle wax		1
13. First aid box (Fully equipped)	portable	1
14. Litmus paper/indicator	Red and blue books	1
15. Phenolphthalein		1

16. Fire extinguisher/sand bucket
   CO_2 type
   Powder type
   Sand
   Blankets
	Cylinders
	Cylinders
	Buckets
	Pieces
		2
		1
		10
		20

17. Roll of cotton wool	Rolls (400g)	1
18. Slides	Microscope slides (76x26mm); permanent slides	1 box 20
19. Coverslips	18mm square	1 box
20. Hand lens	Double x 6 and x20	24
21. Bio-viewer		6
22. Models/charts	Eye, Ear, heart, skeleton	5 each
Essential chemicals		
23. Benedicts' reagent	2 litres	1
24. Sudan dye	Sudan III 0.5 litres	1
25. Eosin powder	50g	1
26. Iodine	500g	1
27. Hydrogen Peroxide	2.5 litres	1
28. Diethyl ether	2.5 litres	1

29. Formalin	2.5 litres	1
30. Yeast	100g	1

31. Sugars
  Fructose
  Lactose
  Sucrose
	
	500g
	500g
	500g
		
		1
		1
		1

32. Enzymes	Diastase 100g Pepsin 100g Trypsin 100g	1 1 1
33. Agar powder	250g	1
34. Cobalt chloride anhydrous	100g	1
35. Lanolin	500g	1
36. Methylene blue	25g	1
37. Chloroform	2.5 litres	1
38. Starch soluble	500g	1
39. DCPIP	5g	1
40. Bench acids	Sulphuric acid 2.5 litres Nitric acid 2.5 litres Hydrochloric acid 2.5 litres	1 1 1
41. Alkalis	Sodium hydroxide 2.5 litres Potassium hydroxide 2.5 litres Calcium hydroxide 2.5 litres Ammonium hydroxide 2.5 litres	1 1 1 1
42. Bicarbonate indicator	500ml	1
43. Universal indicator BDH 1-14	Packet	1
44. Methylated spirit	2.5 litres	1
45. Ammonia solution	500ml	1
Apparatus		
46. Quadrats	1m2	10
47. White tile	pieces	10
48. Potometer	pieces	2
49. Dissection pins	Packet	2
50. Glass rods	Dozen	2
51. Droppers	box	2
52. Boiling tubes	Dozen	1
53. Ballons	Packets	1
54. Bell jar	units	1
55. Beakers	50 ml 100ml 250 ml 500ml 1000 ml	10 10 10 10 2
56. Conical flasks	250 ml 500 ml 1000 ml 2000 ml	1 1 1 1

| 57. An assortment of empty bottles, tins, and cans | Collect kitchen, nearby cafes, etc. | several |
| 58. Potassium iodide | 500g | 1 |

School Equipment Production Unit (SEPU) Biology Kit

The SEPU is a Kenya Government undertaking managed by the ministry of education. The SEPU biology kit is a wooden box in which some special apparatus and materials are stored and used when required for experimentation and demonstration work. SEPU has prepared kits in biology, chemistry, physics and primary science to facilitate teaching and learning in these areas. The kits are meant for use during the practical. The SEPU biology kit contains 30 different items of apparatus and materials. The items are as indicated in Table 11.2.

Table 11.2: Items of Apparatus and Materials contained in SEPU Biology Kit

ITEM	QUANTITY
Pegboard strip	1
Pegboard strip for moments	1
Screw, short	1
Knurled nut	1
Mass 20g	2
Mass 10g	1
Beaker 400ml	2
Test tube stand	1
Dropper	2
Thermometer	1
Hand lens	1
Viewer	1
Set of slides (13 slides in each)	1
Test-tube, large	1
Test-tube, disposable	6
Test-tube stand base	1
Filter Funnel, plastic	1
Petri dish	2
Syringe 10ml	1
Cork stopper	4
Mounted needle	1
Forceps	1
Scalpel handle	1
Blades for scalpel	2
Root marker	1
Modelling clay half packet	1
Wooden block	4
Scissors	1
Visking tubing 15 cm	2

Advantages of Using SEPU Biology Kits

SEPU Biology kits have several advantages:

- The kits are portable and the teacher can move with them from one class to another as well as outdoors in the field;
- The kits are made using locally available materials and are therefore generally cheap;
- The kits encourage learning by doing as the students become actively involved in handling and doing experiments;
- The kits offer teachers some authentic contact with practical biology into the classroom and can inspire teachers to try out new ideas and improvise on new experiments (innovation); and
- The kits are useful for teaching and learning certain specific facts and principles through demonstration and experimentation and include a full set of materials for teaching some difficult concepts in the syllabus.

Improvisation of Resources in Biology

Improvisation is the use of low-cost materials drawn from the environment to make teaching resources and may include modification of apparatus or equipment for that purpose. The biology teacher is encouraged to improvise whenever a deficit exists in the provision of apparatus needed for conducting practical. Some of the improvisations include:

- Construction of quadrats for ecological studies using wires or wood fastened with nails;
- Make funnels by cutting empty plastic bottles using a hot wire or a sharp knife. The base of the remaining bottle can be used as test tube holders;
- Mortar and pestle can be made from a stump of wood. The mortar is made by scooping out pieces of wood from a stump, while the pestle is made by curving out from a piece of straight wood. You can also purchase a cheap wooden mortar and pestle made locally by carpenters;
- Modelling clay may be used to make models of organs such as ear, heart and eye;

- Construction of a ribcage model using cartons or strips of cardboard obtained from a carpenter. Use threads or nails for fastening the joints;
- Graduated plastic containers cheaply purchased from shops can be used as measuring cylinders;
- Cheesecloth may be used as filters;
- Tin lamps may be used as sources of heat instead of Bunsen burners;
- Plastic containers including medicine caps may be used as Petri dishes; and
- Drinking straws cheaply purchased from shops may be used for capillary experiments.

Preparation of Solutions

Some of the common solutions you will require for experimental work in the biology laboratory are prepared as outlined in Table 11.3.

Table 11.3: Preparation of Commonly used Solutions in the Biology Laboratory

Solution	Preparation
Potassium iodide	Dissolve 20g of potassium iodide and makeup 1 litre
Iodine solution	Add iodine crystals to an aqueous solution of potassium iodide until saturated. Filter and dilute to obtain a pale golden brown solution.
Benedict's solution	Dissolve 17.3 g copper sulphate, 173g sodium citrate, 100g anhydrous sodium carbonate in water to make up to 1 litre.
Methylene blue	Dissolve 1g of methylene blue in 100ml of distilled water. Add 0.5 g sodium chloride (for use on living structures). Dissolve 0.3g methylene blue in 30ml 95% ethanol to which 100ml of distilled water has been added(for use on non-living structures)
10% sodium hydroxide	Dissolve 100g sodium hydroxide in a litre of water.
1% copper sulphate	Dissolve 10g of copper sulphate in a litre of water
Dil. Hydrochloric acid	Add 172 ml concentrated acid to water to make up to 1 litre (2HCl)
Cobalt chloride paper	Immerse sheets of filter paper in a 5% aqueous cobalt chloride solution. Remove and blot. Dry the papers in an oven at 40 degrees Celsius. Preserve in a desiccator containing anhydrous calcium chloride.
Formalin	Formalin contains about 40% methanol or formaldehyde. Before use, add 90 ml of formalin to make 10% formalin which is a fixative.
Dichlorophenolindophenol (DCPIP)	Dissolve 50 g of DCPIP powder in 50ml water to make 0.1 % solution.

Preservation of Specimens

The materials or specimens (plants and animals) collected from the field should be preserved in a state that retains their life characteristics as much as possible.

Preservation of plants

Plants are best preserved for use in the **herbarium**. A herbarium is a collection of preserved plants either as entire plants or as plant parts such as flowers, root, fruits and leaves. The plants are first placed between absorbent papers such as newspapers. The paper is folded back and placed in a warm, dry and well-aerated place for about five days. Heavyweight (such as several large books) is placed on top of the paper containing the plants. Once dry, the plant is mounted on Manila paper using cellotape and details are written on it at the lower left corner. The details include scientific name of the plant; locality it was collected from; altitude; habitat; description; use; date collected; and collector's identity. The mounted specimen can be protected by applying a solution of mercury chloride in alcohol to prevent it from destruction by pests and fungi.

Leaves mounted on herbarium papers. Photo/ Fundación Cerezales/FLICKR

Preservation of animals

The animals commonly preserved in the biology laboratories include, but not limited, to the following: Fish; Earthworms; Reptiles; Centipedes and millipedes; Insects; Grasshoppers; Birds; and Bats. The animals are usually collected from their habitats and then anaesthetised by adding suitable chemicals such as 70% alcohol or chloroform. They are then preserved appropriately. Table 11.4 gives a summary of the process.

Table 11.4: Methods of Preserving Animal Specimens in the Biology Laboratory

Animal	Killing	Preservation
Fish	Anaesthetised using chloroform	10% formalin
Earthworm	Anaesthetised using chloroform	5 % formalin
Crustaceans (e.g. Crayfish)	Immerse in formalin	8% formalin
Reptiles (e.g. snakes)	Injecting with 70% alcohol	
Centipedes and millipedes	Anaesthetised using chloroform	Carl's solution(prepared by adding 170ml of 95% alcohol to 60ml of 40% formalin and 280ml of distilled water, with a small quantity of glycerine.
Insects	In fumes of chloroform	70% alcohol
Birds	Anaesthetised using chloroform	70% alcohol
Bats	Anaesthetised using chloroform	70% alcohol

Preparation of Skeleton Specimens

Specimens of small animals, as well as large herbivores, carnivores and rodents, may be collected from the field and then skeletons are prepared. The skulls and teeth of such animals may also be collected separately in the field. Some bone specimens may be obtained from butcheries and hotel remains. In addition, the dissected rabbits and rats may also be treated for the purpose of obtaining bone specimens.

In case of preparing fresh skeletons the following procedures are used:

- Bury the specimen in the soil for several weeks to decompose them;
- Unearth them and brush off the soil carefully;
- Soak the specimen in a strong detergent for several days, changing the solution periodically;
- Dry the specimen in the sun; and
- Label the bone specimen using a marker.

The Live Specimens in the School

The School Aquarium

An aquarium is a vivarium of any size having at least one transparent side in which aquatic plants or animals are kept and displayed. The school aquarium is usually small and made from locally available materials. The rectangular frame is made of metal and glass fitted on the three sides except for the top side which is left open.

A school aquarium. Photo/PAUL KOBAYASHI/FLICKR

The frame may also be made of softwood fitted by a carpenter around a polythene sheet or a high strength acrylic material. The facility is then filled with fresh water and some clean sand at the base. The sand has a filtering role. In a commercial aquarium, there is usually specialised equipment for aerating and filtering the system.

The animals commonly used are freshwater fish, invertebrates, amphibians, and aquatic reptiles such as turtles. The plants include Eldoa and Spirogyra. The plants add oxygen to the water through photosynthesis to support the gaseous exchange and respiration in animals. In turn, the animals supply carbon dioxide needed by the plants during photosynthesis. However, animals should not grow too large.

The aquarium may be used by students in studying locomotion in fish, gaseous exchange in plants and animals and ecological relationships.

The School Egg Incubator

The school egg incubator is a device used to simulate avian incubation by keeping eggs warm through maintenance of optimal temperature, humidity and other conditions such as turning the eggs for them to hatch. The type of egg incubator required for use by students is a small one and usually cheap. It may be powered by solar or electric energy as the need may be. The egg incubator is used when the students are required to conduct projects on the stages of development of chicken

The School Garden

A piece of a plot measuring about 20mx20m within the school compound or negotiated out with members of the surrounding community will be useful for the teaching of biology. It can be used for practical activities including projects such as investigating the effect of various type of fertilizers on plant growth, stages in the germination of seeds, and soil types, among other issues.

Biology Text Books

Textbooks are an important source of information for students and if well used can help them develop the requisite skills and attitudes such as study skills and personal research. It is therefore important to know how to select the best text and reference books for them. The following are some of the criteria that can be used in selecting the books:

An egg incubator. Photo/DEEPAK GOLA/FLICKR

- The content should be related to the relevant general curriculum in terms of the child's environment and the country's socio-economic needs and specific subject objectives;
- The content should contain up-to-date ideas and information in the subject;
- The content provided should be adequate and of required depth and breadth;
- The organisational structure and layout of the book and the sequencing of the topics and sub-topics should be progressive and organized in such a way as to allow maximum learning by the students;
- The content should be supported by adequate and aesthetic illustrations, pictures, diagrams and photographs to arouse curiosity and interest in the learners so that they want to read and learn more

about the content in the book. The use of appropriate colour and familiar environments is usually appealing to them;
- The print used should be appropriate (font large enough to be read without difficulty) and the design should be interesting to the learner;
- The language used should not be vulgar and should be appropriate to the age of the learners;
- The resources suggested should be easily attainable and should motivate and interest the learners;
- The covers and binding should be durable;
- The cover prices should be moderate and affordable by most parents and the nation at large;
- The activities suggested should enhance scientific investigations and development of inquiry skills; and
- The evaluation/assessment procedures suggested and tasks suggested should be appropriate to the mental development of the learners and be related to their required knowledge and skills

Teaching with Information and Communication Technology (ICT)

Computer technology is the key driver of ICT. It has potentially many applications in the teaching of biology. The opportunities include, but not limited to the following:

- An Internet search for information;
- Video clips of biology lessons;
- Worksheets and practical instructions;
- Pictures and illustrations of abstract concepts and processes;
- Word processing;
- Websites for information;
- Animated presentation of abstract biological concepts and processes;
- Data analysis;
- Statistical analysis;
- Graphic presentation of information; and
- PowerPoint presentation of papers.

The computer technology has several opportunities for the teaching and learning of biology because of two key processes associated with it, namely, **scaffolding** and **affordances**.

Affordances

The term affordance refers to the range of potentialities offered by ICT tools for the purpose of learning. The affordances offered by the computer are wide-ranging and include viewing animated illustrations of the otherwise abstract concepts like germination of seeds, plant growth and development, physiological processes like osmosis and plasmolysis, complicated structures and processes like DNA and protein synthesis among others. Affordances are therefore about properties that provide opportunities for action. The opportunities offered by ICT, and computers, in particular, shape and urge particular uses (Allen and Otto, 1995; Gibson, 1979). This means that the manner in which ICT provides a wide range of possibilities depending on the design of the application and the choices offered.

Vera and Simon (1993) see affordances as a set of cognitive constructs occasioned by ICT activities that bring about some change in mental representation resulting in new mental structures.

The students' ICT proficiency will clearly affect the possibilities of using ICT. Pupils' capability with ICT will also affect their capability for independent learning, their meta-cognitive skills, their own perceptions of the learning objectives for a particular lesson and their perceptions of the value of ICT for their learning (Newton and Rogers, 2003).

It is necessary for teachers in their planning to consider how they expect ICT to support, enhance or transform teaching. This means identifying the features that will provide affordances, and evidence of the affordances in practice. The affordances provided by ICT should be distinct from that provided by the teacher, the other pupils and other resources. However, the relative balance of these and their interrelationships should be taken into consideration in the entire learning process. For example, peer support may direct pupils towards features of ICT which may then become affordances (John and Sutherland, 2005). The learning outcomes are thus a product of the affordances and learning activities and the interaction between them. Thus teachers need to plan appropriately in order to enable pupils to engage in challenging ICT-based activities. Important still, students should be equipped with the basic skills to work with computers.

Scaffolding

The major task of the biology teacher is to organise and support the students in their investigation or inquiry for the purpose of seeking evidence for answering the questions they raise. The provision for the assisted performance of the students by the teacher is known as scaffolding. The teacher provides assistance to the students with the aim of achieving the expected learning outcomes (ELOs) in terms of the learners' performance. The assistance is gradually withdrawn so the learner can perform the tasks independently.

The term 'scaffolding' comes from the works of Wood, Bruner and Ross (1976). The term was developed as a metaphor to describe the type of assistance offered by a teacher or peer to support learning. The meta-phor of a scaffold captures the idea of adjustable and temporary support that can be removed when it is no longer needed (Vaughn and Bos, 2009). Scaffolding today refers to the action of adjusting and extending instruction by the teacher so that the student is challenged and able to develop new skills and knowledge. Scaffolding is actually an enabling bridge upon which students build on what they already know to arrive at something they do not know (Benson, 1997).

Multimedia interactive ICT increases the possibilities of the assistance since it allows for direct as well as indirect instructional pathways that expand problem-solving situations thus making learning possible. The students learn in the process because they internalise the knowledge transacted through assisted performance, while the teacher's task is to help the students overcome any barriers to the learning of the tasks in question. The key elements of scaffolding include:

- Task definition;
- Direct or indirect instruction;
- Specification and sequencing of activities;
- Provision of materials, equipment and facilities; and
- Other environmental contributions.

In the process of scaffolding, the teacher helps the student to master a task or concept that the student is initially unable to learn or perform independently. However, the teacher provides assistance with only those skills that are beyond the student's capability. When the student takes responsibility for or masters the task, the teacher begins the process of withdrawing the scaffolding, which allows the student to work independently.

The teacher can scaffold instruction to meet the needs of the students by manipulating the task, materials, group size, pace, presentation, among other things. Some of the tools that can be used in scaffolding students' learning of biology include:

- Breaking the task into smaller more, manageable parts;
- Using 'think aloud', or verbalising thinking processes when completing a task;
- Cooperative learning, which promotes teamwork and dialogue among peers;
- Concrete prompts, questioning;
- Coaching;
- Cue cards or modelling;
- Activation of background knowledge;
- Giving tips, strategies, cues and procedures;
- Minimising students' stress level by monitoring and assisting them;
- Teaching students in small segments by using explicit, systematic instruction such as modelling, guided and independent practice, and use of consistent instructional procedures. Systematic instruction refers to sequencing instruction from easier to more difficult ideas and teaching the easier skills to mastery before introducing more complex skills; and
- Understanding the student's prior knowledge and abilities. Prior knowledge allows the student to be connected to the new knowledge and made relevant to the learner's experiences, thus increasing the motivation to learn.

There are five different methods in instructional scaffolding: modelling of desired behaviours, offering explanations, inviting students to participate, verifying and clarifying student understandings, and inviting students to contribute clues (Lange, 2002). These techniques are used to direct students toward self-regulation and independence.

Generally, scaffolding may include assistance with planning, organising, doing and/or reflecting on the specific task. Such assistance is best made available in a timely manner matched to the learning needs and interests of the learner. Scaffold instruction may include the following steps:

- Break the task into small steps;
- Teach easier skills first, then more difficult ones next;
- Slow the pace of new skill introduction to allow for more practice of a task;

- Use a small group size;
- Make thought process for accomplishing tasks overt by talking to students about what you are thinking when you engage in the talk. Have them share what they are thinking when they practice the task;
- Teach strategies for completing complex skills;
- Model all steps involved in completing tasks;
- Provide assistance during the first student attempts at skills;
- Praise the accomplishments of each small step;
- Use concrete materials during initial skill instruction; and
- Vary the materials used.

It is apparent that scaffolding is significant in the learning process as it makes the following contributions:

- It makes it easier for the learner to undertake a task successfully;
- It expands the possible learning activities and experiences;
- It increases the rate at which learning may be achieved;
- It extends what is possible for a learner to perform;
- The provision of powerful tools and well-formed instructions enable higher order problems to be solved more rapidly;
- Possible early identifier of giftedness;
- Provides individualised instruction;
- Greater assurance of the learner acquiring the desired skill, knowledge or ability;
- Efficiency in delivery, since the work is structured, focused, and glitches have been reduced or eliminated prior to initiation, time on task is increased and efficiency in completing the activity is increased;
- Creates momentum – Through the structure provided by scaffolding, students spend less time searching and more time on learning and discovering resulting in quicker learning;
- Engages the learner;
- Motivates the learner to learn; and
- Minimises the level of frustration for the learner.

The challenges of using scaffolding instruction:

- Lack of proper computer infrastructure in schools;

- Lack of adequate knowledge, skills and attitudes on the part of the teachers to be able to use ICT to scaffold learning activities in the biology classes;
- Curriculum materials may have to be designed, developed, and provided with the necessary scaffolding for teachers to easily customise them for the short-term needs of the learners;
- Very time consuming;
- Inadequately modelling the desired behaviours, strategies or activities because the teacher has not fully considered the individual student's needs, predilections, interests, and abilities (such as not showing a student how to 'double click' on an icon when using a computer);
- Full benefits not seen unless the instructors are properly trained;
- Requires the teacher to give up control as fading occurs; and
- Lack of specific examples and tips in teacher's editions of textbooks.

Box 11.1

Discussion

1. List as many websites as possible that are a good source of teaching and learning resources in biology

2. Give your students an assignment on the helical structure of DNA. Ask them to explain the structure. They may visit a website provided by you.

Conclusion

This chapter has established that the learning of biology is best achieved if the teacher supports the students' inquiry or thinking through the provision of instructional resources. In the process, the student seeks and attains 'the truth' about the problem that was driving the learning process. There are many types of such instructional resources ranging from real living things to simple materials in the environment to digital resources. It is the responsibility of the biology teacher to equip, prepare and guide the students effectively during the learning process. This chapter has outlined the rationale for the use of instructional resources and provided information on the varying types of such resources ranging from realia to simple materials to digital resources to facilitate the learning of biology.

References

Allen, B. & Otto, D. (1995) *Media as lived Environments: The Ecological Psychology of Educational Technology*, in: D. H. Jonassen (Ed.) Handbook of Research in Instructional Technology (New York, MacMillan).

Benson, B. (1997). Scaffolding (Coming to Terms). *English Journal, 86(7)*, 126-127.

Cox, M. and Webb, M. (2004*) An Investigation of the Research Evidence Relating to ICT Pedagogy*. Becta, ICT Research. (http://www.becta.org.uk)

Ellis, E. (1994). *Effective Teaching Principles and the Design of Quality Tools for Educators*. Retrieved March 15, 2004, from http://uoregon.edu/~ncite/documents/techrep/tech06.html.

Ellis, E. S., Worthington, L., & Larkin, M. J. (n.d.). Executive summary of the research synthesis on effective teaching principles and the design of quality tools for educators. Retrieved March 25, 2004, from http://idea.uoregon.edu/~ncite/documents/techrep/tech06.html.

Galloway, C. (2001). *Vygotsky's Constructionism*. Retrieved January 20, 2004 from http://projects.coe.uga.edu/epltt/index.php?title=Vygotsky%27s_constructivism.

Hogan, K., & Pressley, M. (Eds.). (1997). *Scaffolding Student Learning: Instructional Approaches and Issues*. Cambridge, MA: Brookline.

Lange, V. L. (2002). *Instructional Scaffolding*. Retrieved on September 25, 2007 from http://condor.admin.ccny.cuny.edu/~group4/Cano/Cano%20Paper.doc

Vaughn, S.and Bos,C.(2009). *Strategies for Teaching Students with Learning and Behavior Problems*. 7th edn. New Jersey, Person.

Vera, A. H. & Simon, H. (1993) *Situated Action: A Symbolic Interpretation, Cognitive Science*, 17, 177–186.

Watson, D. M. (2001) *Pedagogy Before Technology: Re-thinking the Relationship between ICT and Teaching, Education and Information Technologies*, 6(4), 251–266.

Wood, D., Bruner, J.S., & Ross, G. (1976). The role of Tutoring in Problem Solving. *Journal of Psychology and Psychiatry*. 17.

TEACHING SECONDARY SCHOOL BIOLOGY

GROUNDING IN pedagogy is a prerequisite to developing effective teaching methodologies

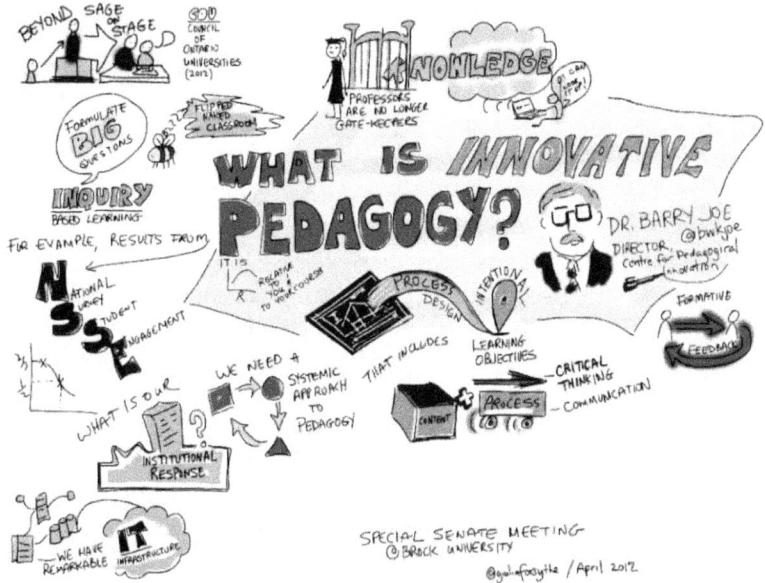

Photo/GIULIA FORSYTHE/FLICKR

12

CHAPTER TWELVE

TEACHING BIOLOGY: APPRAISING PRACTICE

◆ Teaching Skills ◆ Self – Appraisal in Biology Instruction ◆ The 21st Century Pedagogy ◆ Transformative Learning ◆ Appraising Learning in Biology

Introduction

Improving the quality of biology teaching at the secondary school level has been a legitimate concern among the various stakeholders including teachers, teacher educators, education authorities, sponsors, teachers' professional organisations, among others. The need to improve the professional training and performance of teachers is now widely acknowledged.

Grounding in pedagogy is a prerequisite to developing effective teaching methodologies. For example, the demands of education at all levels today call for teachers who are well skilled and grounded in knowledge, values and teaching strategies which facilitate the creation of an enabling teaching and learning environment (Quist, 2000). To provide this, teachers need to demonstrate competency in the various teaching skills such as planning and organising lessons, teaching techniques and strategies, using computers in lessons, working with others, assessing pupils activities, reviewing their own approaches, and selecting and using suitable resources to support the teaching and learning processes. As teachers, we must be able to understand these dimensions and appraise our practice with a view to improving our approaches and practices.

This chapter is intended to outline the most basic competencies for teaching biology and the mechanisms for appraising our teaching for the purpose of improving our practice.

Teaching Skills

Teachers are professionals and must exhibit high levels of teaching skills. These are the skills that facilitate learning; they improve with practice, most teachers develop the skills as they work in their classroom. They include the five most basic skills that teachers expressly use in every lesson, namely:

- Set induction;
- Questioning;
- Reinforcement;
- Using examples;
- Feedback;
- Monitoring and Evaluation; and
- Closure.

Other skills include:

- The organisation of the classroom;
- Planning and preparation;
- Classroom management;
- Effective communication; and
- Personal Practical Knowledge or Professional responsibilities

Set Induction

This is usually referred to as an introduction to the lesson. It sets the stage for the lesson and usually takes about 5 minutes. The mind of the learners must be prepared during the introduction to receive knowledge. The introduction should not only set the atmosphere or need for learning but it should gather together the students' attention and focus it on the lesson. The functions of introductory activities are therefore to:

- Help pupils gain mental alertness and preparedness for the lesson;
- Link the lesson with previous knowledge;
- Spell out what the topic is all about;
- Check on the relevant knowledge the students already have on the topic; and
- Help arouse students' interest and curiosity in the lesson (raise pupils expectations in the lesson).

The introduction activities include:

- Testing of previous knowledge and degree of receptivity through oral questioning, a demonstration, test, etc.; and
- Arousing interest and curiosity through activities such a demonstration, storytelling, displaying a chart, showing a model, giving a puzzle or problem to be solved.

Closure

Conclusion or closure refers to actions or statements by teachers that are designed to help pupils bring things together in their minds, to make sense out of what has been taught during the lesson (Shostak, 2003). Research shows that learning increases when teachers make a conscious effort to help pupils organise the information presented to them and perceive relationships based on that information (Shostak, 2003). The functions of closure are to:

- Help draw students' attention to a closing point in the lesson;
- Help students consolidate their learning – i.e. tying all the information or activities into a meaningful whole;
- Reinforce the major points to be learned so that the students are helped to retain the important information learned in the lesson;
- Relate the lesson to the original organising concept or principle or theme;
- Lead students to extend or develop new knowledge from previously learned concept or principle; and
- Allow students to practice what they have learned.

The lesson can be concluded through a variety of activities, including:

- Providing a summary of the main aspects of the lesson;
- Taking notes, for example in the form of filling in the blanks;
- Giving homework or assignment;
- Oral questioning on main aspects of the lesson;
- Giving a short objective type test;
- Asking pupils to label sketches or draining;
- Organise thinking around a new concept or principles; and
- Follow – up to the film, T.V programme or Video show.

Organisation

This includes: organising the classroom to make the best use of space; providing clear rules and routines; keeping pupils working actively; safety and arrangement of furniture.

Planning

This involves – structuring the content to suit the learners; making sure that the syllabus is properly covered; thinking about the best way to present the lesson; making provision for children working at different cognitive levels; providing a variety of activities e.g. listening, oral, written and practical work; setting realistic goals and objectives, designing suitable instructional materials and resources; procedures for assessing students' learning.

Classroom Management

This deals with class control; time management; the management of both group and individual learning; management of teaching resources by making the most of local materials; and teacher interaction with pupils.

Communication or Instruction

This involves – **questioning** and explaining new ideas; relationship with the learners; instructions; **using examples** to facilitate understanding; relationship with other teachers; relationship with parents and the school management; **scaffolding learning**; providing feedback; and responsiveness to students.

Monitoring

This focuses on observation and assessment of pupils learning progress. Observation deals with more than one thing at once – knowing exactly what is going on in the class even when your back is turned or you are attending to the needs of another pupil – i.e. to ensure that everyone is paying attention. Assessing progress involves judging the progress and learning achievement of all pupils – do all the pupils understand what needs to be done? Is anyone having difficulties?

Evaluation

This involves evaluating the learning achievement of pupils through questions, written work and tests and practising self-reflection (the ability to honestly evaluate the strengths and weaknesses of your own teaching).

Personal Practical Knowledge or Professional Responsibilities

This includes the set of understandings teachers have of the practical circumstances in which they work such as beliefs, insights, and habits that enable them to do their work effectively. These include:

- Growing and developing professionally – e.g. enhancement of content knowledge and pedagogic skill;
- Developing one's instructional philosophy in biology;
- Contributing to school and community projects;
- Maintaining accurate records;
- Communication with parents, colleagues, school and administration;
- Service to students, community, and nation;
- Gender stereotypes – shared beliefs about the work roles or behaviour of men and women; for example "Men are better workers that women" or "A woman's place is in the home";
- Gender bias – attitudes and practices which show gender stereotypes – e.g. showing one gender, either boys or girls, as always in the lead, active, successful and valued;
- Gender fairness – believing that boys and girls have equal value and making sure that they have equal opportunities to learn and to take responsibility;
- Reflection – a process of self - appraisal or self-evaluation that allows you 'to think above your teaching skills and practice – e.g. evaluating strengths and weaknesses of your teaching (SWOT analysis). Reflection should be an on-going activity, each time making notes on each lesson in your record book. It can be done alone, with colleagues or as part of an appraisal process; and
- Reflection can help improve the quality of your performance if you take time to do it – e.g. what you need in order to improve the content of lessons or the approach to topics; the changes you need to make before giving other lessons on a similar topic; the methods you need to use for better results; improving time management; the kind of support you might need to help you sustain improvement.

Self - Appraisal in Biology Instruction

Quality of Performance in Biology

In biology, the teacher is expected to help students develop scientific skills as well as knowledge. The teacher who emphasises the scientific inquiry rather than the transmission of scientific information faces the problem of how to involve pupils in problem-solving. Some teachers overcome this problem by raising questions and then devise ways of answering them. In doing this, the teachers aim at a difficult compromise between product and process, between covering the syllabus and involving pupils in the kinds of thinking that constitute scientific inquiry. Their dilemma is posed sharply in the following extract of a biology lesson:

Biology Practical Lesson 1 (Extract)

Teacher: Today, you are going to cover the leaves of a growing bean seedling with silver foil with holes cut in it, and later to test the covered and uncovered surfaces for starch.

Jane (*intercepting*): Madam, I don't see how that will prove it. It could be all sorts of other things we don't know anything about.

Teacher (*comes down the laboratory to pupil's 1 bench*): Can you expand your question? Explain what you mean!

Jane: Well, you said if there was starch in the bare patches it would mean that there was----it was because of the light, but it could be the chemicals in the foil or something we know nothing about.

Martha (intercepts): Of course it will prove it; we wouldn't be wasting our time doing the experiment if it didn't.

Teacher: I don't think that's a very good reason----eh—eh—yes Mercy---

Mercy (anxiously interjecting) ----Anyway, we shouldn't waste more time----let's get down to business.

The dilemma is evident: The teacher has to choose between the authoritative knowledge, which is taught with an emphasis on right answers and with little attempt to involve pupils in the kinds of reasoning that scientific inquiry entails, and scientific inquiry that deals with the challenges of the adequacy of experimental design. The weight of each of the two approaches is related to the rewards the pupils expect:

- Biology examinations tend to emphasise on the mastery of knowledge that helps students to gain qualifications and career in science; the student has to align to this version of scientific knowledge, and to the kind of learning that goes with it; and

- Intrinsic motivation in which what matters is doing science the right way as scientists do—the pursuit of evidence for the truth-- rather than just accepting the consensus.

Why are teachers faced with this dilemma? Is transmission of knowledge the most suited way for training scientists for employment or is it because it lends itself more readily to school examinations? Science teaching should promote ways of thinking by the students so that they may think out new problems or relate new knowledge to what they already know. Too much teacher talk and chalk still characterise the learning of scientific knowledge in schools. The teaching of science does not seem to challenge the pupils' ability to construct new knowledge through the scientific process. As biology teachers, we should promote students' ability to make sense of the biological world through discussion and writing, and not cram into their heads an overwhelming amount of other people's biological knowledge.

Characteristics of Teacher Talk

An analysis of teachers' lessons gives us an insight into the characteristics of their talk. Read each of the extracts and identify which statements were made by a teacher. How did you tell when the teacher was talking? Write down a list of characteristics of teacher talk.

Biology Lesson 2 (Extract)

Teacher: Can you tell us what fossils are----- do you think, Njoroge?

Njoroge: Sir, Sir, a long time ago animals---there were animals, and when they died, eh, the rain and wind came over them and then the bodies disappeared and left the shells and that fossils appeared---.

Teacher: Good. Why do you think the bodies disappeared and the shells stayed?

Wafula: Sir, sir they rotted----eh—they are soft---

Teacher: Oh, the bodies got rotten!----And what about the shells?

Jane: Sir, they got harder eh, the clay dried, they made marks in the clay----

Teacher: Right---good

Wafula: The clay dried hard—eh—yeah--

Teacher: Right, OK, thank you. Can anybody add anything to that at all? Like what are fossils then?

Sharon: No sir—eh—Tell about it now--

Teacher: It is a very good description.---Let us not waste more time---now take out your books and take the following notes on evolution---

From the lesson extracts, the following are some characteristics of teacher talk:

- Repetition of students reply;
- Reinforcement;
- Questioning - leading; closed or open questions;
- Drawing student attention;
- Giving instruction;
- Clarifying technical terms;
- Using technical terms;
- Long sentences; and
- Evaluation of students' work through encouragement.

The 21st Century Pedagogy

Literature review indicates that present-day pedagogy not only entails all the above dimensions but also includes several other new elements that must but put into consideration for effective teaching and learning. The 21st Century pedagogy entails the elements summarised in Figure 12.1.

The figure clearly shows that for us to teach using best practice pedagogy, we must put students at the centre. Our curricula and assessments must be inclusive, interdisciplinary and contextual; based on real-world examples. Our students must key participants in the assessment process. We must establish a safe environment for them to not only collaborate in it but also to discuss, reflect and provide feedback. In this case, we must make use of collaborative and project-based learning, using enabling tools and technologies to facilitate this. In the process, we develop in the students key fluencies and higher order thinking skills.

Our tasks, curricula, assessments and learning activities are designed to build on the Lower Order Thinking Skills (LOTS) for the purpose of developing Higher Order Thinking Skills (HOTS). Our teaching must also be inclusive of the different learning styles of our learners, such as visual, kinesthetic, read/write or auditory (VARK) learners to facilitate their full participation in the learning process.

The 21st-century pedagogy incorporates new elements of pedagogy such as reflection, collaboration, fluency, affordances, digital natives, innovative teaching approaches (e.g. problem-solving) and scaffolding, thinking skills and inclusive assessment, among others.

(a) **Reflection**: Teaching is required to encourage the students to conduct self-reflection through self-review and peer review for the purpose of improving their learning and performance;

(b) **Collaboration:** Teaching is required to encourage collaboration. Collaboration refers to a holistic overview of the education process whereby individuals build on the value provided by others and enabling and empowering technologies. It includes students working with other students nationally, regionally or globally, incorporating teamwork using their team skills and interdisciplinary approach. Learners may also seek and work with experts as required. Students and teachers are collaborating across the world and beyond the time constraints of the teaching day.

Collaborative learning helps students by allowing them to focus more on real-world problems as a basis of understanding. It also helps them to clarify understanding and support deeper learning. What they learn in one subject becomes useful in other subjects. Projects become more encompassing in an interdisciplinary manner. The sum of the students' learning becomes greater than the individual aspects taught in isolation. Teachers begin to network, collaborating with their peers in other subjects, to link and bind the learning in their subjects;

(c) **Fluency**: Teaching is required to develop various types of fluencies. Fluency refers to unconscious competency in the use and manipulation of media, technology and information. We need the students to be fluent in the use of technology (technological fluency), in collecting, processing, manipulating and validating information (information fluency) and in using, selecting, viewing and manipulating media (media fluency);

(d) **Affordances:** Teaching is required to use enabling technologies or affordances. Affordances are the properties of a system as perceived by the user which allow certain actions to be performed and which encourage specific types of behaviour (Cox and Webb, 2004). For example, digital media such as SMART boards facilitate learning of abstract concepts in Mathematics through such affordances. Enabling technologies including digital tools to lend themselves to wide applications in interdisciplinary approaches and collaborative mediums;

(e) **The digital natives:** These are collaborative, too. The growth of social networking tools like Facebook, WhatsApp, myspaces, Bebo, is fuelled by digital natives. Vast arrays of collaborative tools are available for the purpose of teaching – e.g. classroom blogs, wikis, and social networks. These tools extend the benefits of affordances;

(f) **Problem-solving skills:** Teaching is required to provide opportunities for students to develop problem-solving skills through the use of innovative teaching approaches such as project-based learning, collaborative learning, cooperative learning among others. These approaches should be contextualised based on real-world problems and interdisciplinary realities and facilitated by the use of suitable technologies;

(g) **Scaffolding:** The use of innovative approaches is required to be supplemented with new ways of facilitating learning, particularly scaffolding. Scaffolding is the learning support where pupils build up knowledge and understanding by linking new concepts to those previously understood through a mental framework of linking concepts (Cox and Webb, 2004). It involves adjusting and extending instruction so that the student is challenged and able to develop new ideas and skills. The teacher can scaffold learning to meet the needs of the students by manipulating the task, materials, group size, pace, and presentation (Vaughn and Candaces, 2009). For example, the teacher may scaffold the learning of the students, building on a basis of low cognitive skills such as recall and comprehension to use and apply to tasks requiring higher cognitive skills such as analysis and evaluation. In the process, the students are empowered to make and create. The use of computer programmes can greatly facilitate all these stages in the learning process.

(h) **Thinking Skills:** Teaching is required to help develop both lower thinking skills and higher thinking skills, the former being prerequisite for the development of the latter. Lower Thinking Skills (LOTS) include remembering and understanding and focus on knowledge acquisition. Higher Thinking Skills (HOTS) include applying, analysing, theorising, evaluating, and creating. These focus more on knowledge deepening and knowledge creation.

(i) **Inclusive Assessment:** Teachers are required to conduct an all-inclusive assessment based on clear goals and objectives, relevant assessment tasks, self and peer assessment reports, and timely and appropriate feedback.

Indeed, it is apparent that new developments in ICT provide very different learning opportunities, and a need to design a newly integrated pedagogy as was discussed earlier. We can now define pedagogy as a theory of teaching that focuses on teacher competencies and/or qualities that influence effective teaching and learning.

Transformative Learning

Transformative learning is the latest innovation in teaching and learning, and schools and other institutions are being encouraged to adopt it in their curricula and organisation. Transformative learning is a dynamic process that places students at the centre of their own active and reflective learning experiences. This approach engages learners in deep learning that enables them to develop a broader, more inclusive perspective that is empowering and inspiring.

This change in their ways of thinking opens up their minds to new possibilities about their lives and their environments. They are likely to become productive, creative, ethical, and engaged citizens and leaders contributing to the intellectual, cultural, economic and social advancement of the communities they serve. Thus the students have the opportunity to develop cognitively as well as soft skills (beyond-disciplinary skills) and knowledge important for success as members of local and global communities.

Transformative learning is achieved through innovative and interactive teaching approaches that allow students to look at their lives as a reflection of what they choose to do. They open up and get more interactive with people. They relate concepts in the classroom to the real world. They are changed as persons, for the better, contributing to the community and developing self-worth. They learn to change the world. Through class projects, they discover that it takes a lot of time, effort, and self-motivation to reap the benefits of transformative learning.

They are able to see their hard work transform into something bigger; something more meaningful than just class projects. Transformative education has an immense potential to change the delivery of education at all levels and move students toward a more complete and universal learning experience. However, it requires immense collaborative and innovative processes for the schools and other institutions to engage in to be able to organise and quantify this dynamic learning paradigm

Appraising Lessons in Biology

In light of the information on 21st-century pedagogy, we need to assess the extent to which the biology teachers are incorporating some of these attributes of a modern teacher. The performance of a biology teacher can be assessed by using a tool that allows the assessor to determine the degree of performance in a given lesson. The assessor makes comments for each item on the assessment tool and awards marks on the Likert scale. The lesson

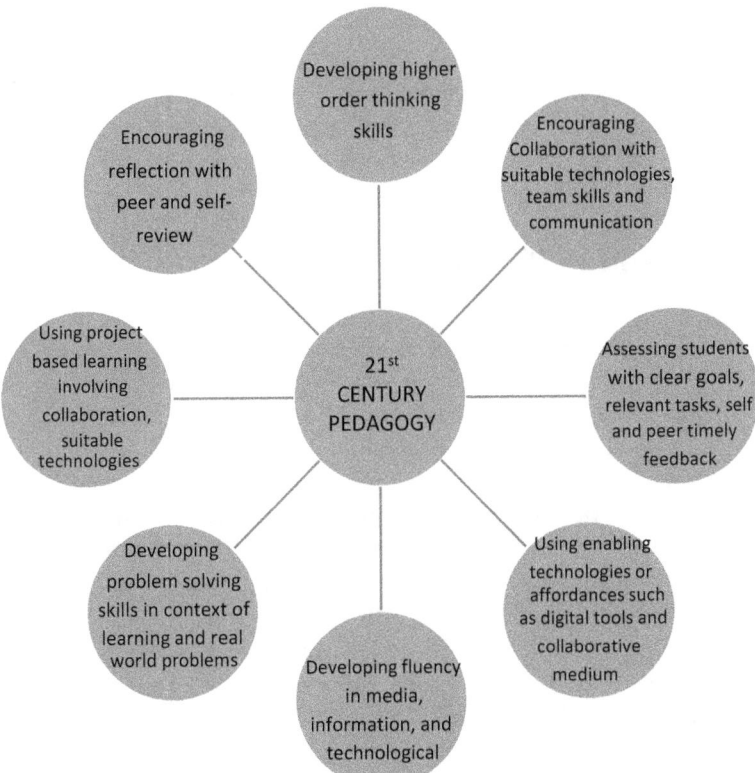

Figure 12.1: 21st Century Skills required for the Learning of Biology

observation and assessment tool contain the following key components:

- Preparation for the lesson;
- Performance; and
- Personality.

The specific items for assessment for each component are as follows:

1) Preparation for the lesson

The areas of focus are as follows:

- **Instructional Objectives**: The objectives should be clearly stated in behavioural terms; they should SMART=Specific, Measurable, Attainable, Realistic, and Time-bound; they should have condition, address an audience, standard of performance expected;
- **Lesson content**: Subject matter in terms of organization, quality, quantity, and relevance to the subject, pupils and learning situation;
- **Teaching aids**: Suitability, quality, relevance, use; and
- **Lesson plan**: Thorough planning indicating the use of time, pupil and teacher activities, content, teaching aids meant to achieve objectives.

2) Lesson Performance

- **Competence in Lesson Sequencing**: Organisation of the lesson in terms of pacing; introduction or set induction(linking lesson to previous one, focusing on pupils' attention, motivating pupils to engage in learning tasks); lesson presentation(involvement of pupils in a sequence of activities, active participation of pupils, use of group work, individual assignments, individual assistance, motivation and focus on individual differences); closure or conclusion(activities to unify the lesson, recapitulation, evaluation and consolidation of achievements);
- **Competence in teaching techniques and approaches**: Appropriateness of procedures and techniques to the subject matter, objectives, pupils' readiness, and teaching situation; effectiveness of teacher's communication, use of pupil feedback, explanation and use of examples, illustrations and demonstration; teacher's supervision, giving of instructions, correcting errors, and consolidating learning; skillfulness in questioning, distribution of questions, and dealing with pupils' questions; Teacher's mastery of content;
- **Competence in use of teaching aids and materials:** Effective use of chalkboard; and quality of writing and set-out; student's quality of records; use of audio-visual aids to promote interest, motivation, communication, understanding, discussion, and thinking skills; use of materials and equipment to promote pupil's task performance well organised; and

- **Competence in classroom control and management:** Teacher – pupil rapport developed during the lesson; Maintenance of good contact with the pupils to sustain pupil participation; maintenance of pupils' interest through the use of lesson materials; Taking prompt action to control pupil discipline; Class was well organised thus well disciplined.

3) The personality of the Teacher

- **Appearance/Image:** Teacher's appearance and presentation in relation to the teaching task/situation; whether teacher's posture and mode of dress inspire the respect and confidence of pupils;
- **Self-confidence:** Mannerism; whether teacher's self-confidence, empathy, pleasantness, firmness promoted teacher-pupil relationship;
- **Communication:** Whether the teacher expressed clarity and adequacy of communication; teacher's voice and language fluency; and
- **Creativity and Innovation:** Adoption of new ways of teaching; improvisation of teaching aids and materials.

The parameters indicated above may be operationalised using a suitable observation tool like the one shown in Table 12.1.

Table 12.1: Biology Lesson Observation Assessment Scheme

Lesson Component	Details of Lesson Component	Key Remarks/ Comments	Excellent (5 marks)	Very Good (4 marks)	Good (3 marks)	Satisfactory (2 marks)	Poor (1 mark)
Lesson Preparation	Objectives						
	Content						
	Teaching resources						
	Lesson plan						
Performance	Lesson sequencing						

	Teaching procedures and techniques							
	Use of teaching aids							
	Classroom control and management							
Personality	Appearance/Image							
	Mannerism							
	Communication							
	Innovation/Creativity							

Conclusion

This chapter has shown that an effective biology teacher must be well grounded in pedagogy. This ensures that biology teachers are well skilled and grounded in knowledge, values and teaching skills which facilitate effective performance. Teachers prepare and demonstrate competency in various teaching skills such as planning and organising lessons, teaching techniques and strategies, using computers in lessons, working with others, assessing pupils' activities, reviewing their own approaches, and selecting and using suitable resources to support the teaching and learning processes. The teachers must also be aware of the latest developments in education including the 21st century pedagogy and transformative learning and incorporate them in their day to day teaching. The biology teachers must understand these dimensions if they have to frequently appraise their practice with a view to improving their approaches and practices.

References

Cox, M. and Webb, M. (2004) *An Investigation of the Research Evidence Relating to ICT Pedagogy*. *Becta*, ICT Research. (http://www.becta.org.uk)

Shulman, L. (1987) 'Knowledge and Teaching: Foundations of the New Reform" *Harvard Educational Review*, 57, pp 1-22

Watkins, C. and Mortimore, P. (1999) "Pedagogy: what do we know?" In: Mortimore, P. (Ed), *Understanding Pedagogy and its Impact on Learning*. London: Chapman.

Vaughn, S and Bos, C (2009), *Strategies for Teaching Students with Learning and Behavior Problems*, 7th Edn. New Jersey: Pearson International.

TEACHING SECONDARY SCHOOL BIOLOGY

A BIOLOGY teacher must be acquainted with the biology laboratory since this is the most fundamental resource in the teacher's work

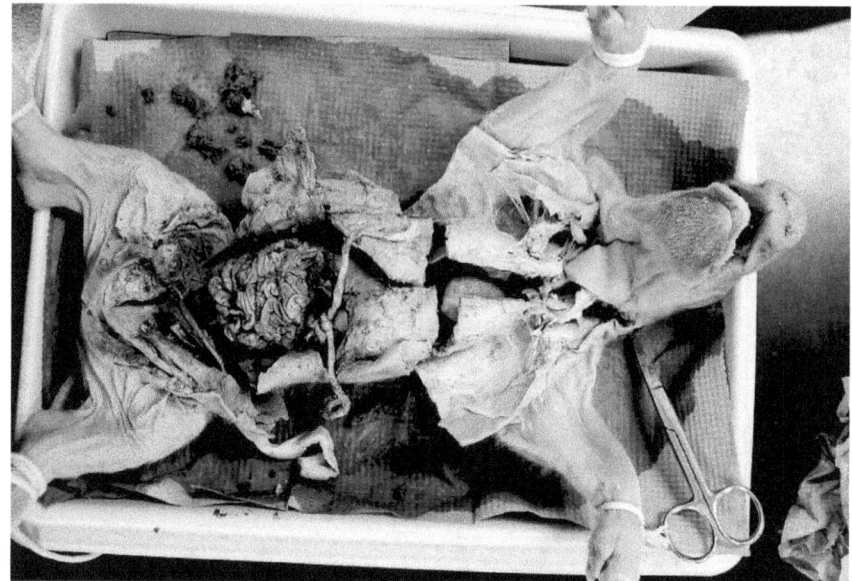

Photo: GLORIA KANG/FLICKR

13

CHAPTER THIRTEEN

THE SCHOOL BIOLOGY LABORATORY

◆ Design and Key Features of School Biology Laboratory ◆ Planning for use of School Biology Laboratory ◆ The School Biology Management Laboratory ◆ Safety in the School Biology Laboratory ◆ School Biology Laboratory Rules and Regulations

Introduction

A laboratory is a place where scientists try to find answers to the questions they ask. The biology laboratory is where the questions regarding living things are investigated and answered. In the laboratory accurate observations using the five senses or extensions of the senses are made. Controlled conditions in the laboratory make it possible to conduct experiments from which conclusions regarding the questions raised are drawn by logical reasoning. Precise measurements and records of the observations are made to aid scientists in arriving at such conclusions. While the school biology laboratory may not be that sophisticated or anywhere comparable to an even modest science research laboratory, the school is expected to provide this facility for the purpose of teaching biology. It is only by experience in the school laboratory that students would appreciate how the knowledge of science has come into existence and why the discipline has shaped human history.

The use of the biology laboratory for the purpose of teaching has many advantages to the students of biology, namely:

- Development of scientific or practical skills;
- Acquaintance with the scientific way of working to solve problems;
- Promoting an understanding of scientific concepts that are otherwise abstract in nature; and
- Developing requisite scientific attitudes that promote positive orientation towards the learning of biology. For example, colourful charts of biological concepts posted on the walls can stimulate students' interest in biology.

The biology teacher must be acquainted with the biology laboratory since this is the most fundamental resource in the teacher's work. This chapter provides information on the design of the biology laboratory based on the more recent standards and guidelines that emphasise more on safety and flexibility, and changes in technology, curriculum and increased student enrolment in the subject.

The Design and Key Features of the School Biology Laboratory

The design of the biology laboratory is premised on four considerations:

- Provision of ample working space for students to perform investigations;
- Ensuring that flexibility of the various sections of the laboratory to accommodate any changes deemed necessary for learning;
- Ensuring that safety measures override other needs such as flexibility; and
- Cost-effectiveness of the entire laboratory set up.

The size of the biology laboratory depends on the needs and financial disposition of the individual school. However, a medium sized laboratory would be equivalent to the size of a classroom accommodating 50 students. The preferred orientation of the laboratory is the East-West, in which the windows are on the North-South sides of the laboratory. This orientation would help avoid too much sunlight into the laboratory that would have an adverse effect on materials and chemicals especially the photosensitive ones like silver nitrate, nitric acid and potassium iodide. The shutters should open outwards to make an easier exit of students in case of an emergency requiring them to run for safety.

The most ideal biology laboratory consists of three sections, namely, the main working space, the preparation room and the store (see Figure 13.1). The main laboratory room is usually the largest of the three. This is where the teacher conducts demonstrations and the students conduct their experiments. It is fitted with benches (worktops) and provided with stools or high chairs on which students sit. The benches may be either fixed (fixed bench design) or adjustable or movable benches. In the former, the benches are provided with working facilities such as Bunsen burners with gas supply, water taps with running water, electricity points and reagents in bottles. In the latter arrangement, only the teacher's demonstration bench or table is fixed. The side benches are provided with working facilities such as water, sink, gas taps, electricity points, and reagents. The latter design is more flexible and is commonly preferred in many schools to allow teachers the option of creating more space (by pushing the tables around) for students to work in groups around the side benches or to participate in the teacher's demonstrations from the teacher's bench to the front. The design is also applicable where the teacher uses the laboratory for classes that don't need practical work. Just behind the teacher's bench or table along the partitioning wall is usually fitted the chalkboard or sometimes the whiteboard for writing as the case may be for writing. On the sides of the main board are usually fitted soft boards for attaching or fastening charts, assignments and general laboratory information. The upper side walls well above the benches are sometimes fitted with lockable glass cupboards in which resources are stored including specimens (both preserved and dried), models, skeletons, herbaria, commercial charts, realia and other resources.

The preparation room is where the teacher and the laboratory technician try out the experiments before being conducted by the students. The room may also be used for planning and keeping laboratory records, laboratory manuals, reference books, worksheets, and chemical ledgers.

The storeroom is used for storing chemicals, equipment and apparatus for future use. The chemicals are usually kept according to the nomenclature and reactivity. Some of the chemicals and apparatus may usually be kept in the side cupboards that are lockable.

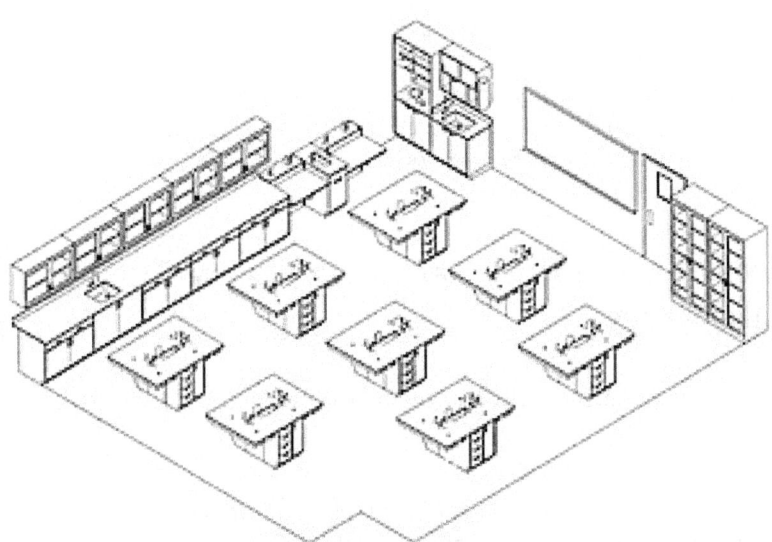

Biology lab layout. Illustration by WBDG https://bit.ly/2VcrgN9

The gas is distributed to the Bunsen burners through gas tubing from the gas cylinder which is usually fitted in a special enclosure on the side of the wall outside the laboratory. On the other hand, the fume chamber or fume cupboard is usually located behind the teacher's bench somewhere between the preparation and the storerooms. The chamber is used to house some experiments involving emanation of dangerous gas whenever such an experiment is conducted or for storage of volatile and smelly reagents. The fume cupboard is usually fitted with a pump, a gas tap, a sink and water, an electricity point, and a sliding shutter.

Planning for use of School Biology Laboratory

Time Table and Requisitions

In many schools, there is only one or two multipurpose laboratories. Laboratory facilities are usually in short supply in such instances. Proper planning is usually required and a timetable for laboratory use must be worked out and displayed in the laboratory. The timetable should indicate which subject, class and teacher will use the laboratory for the double lessons in the course of the week. Even where there is a separate biology laboratory, a timetable should be worked out and displayed appropriately.

In schools where the biology teacher is supported by a laboratory technician, the teacher should usually make prior arrangements with the technician to ensure the apparatus, equipment and materials required for the practical are availed before the class commences. The booking should be made in a requisition or preparation book as the case may be at least one week in advance.

Role of Laboratory Technician in the Biology Laboratory

The technician should attend to the following things:

- Keep the laboratory benches clean and tidy;
- Clean and store glassware and other apparatuses;
- Keep apparatus, equipment and glassware in the correct storage places in the laboratory;
- Check on timetables for the use of the laboratory;
- Assist the biology teacher in the day-to-day preparation for class experiments or teacher demonstrations;
- Pre-test the experiments;
- Keep an inventory of all apparatus, chemicals and equipment;
- Prepare stock solutions for use;
- Set out the laboratory for practical classes;
- Set up apparatus for practical work;

- Assist the teacher as much as possible especially on some technical aspects of the apparatus and equipment;
- Maintain apparatus and equipment in working conditions;
- Prepare visual aids and materials for use by the class;
- Conduct stock taking and make recommendations pertaining to the purchase of any other apparatus and equipment;
- Opening and closing the laboratory; and
- Making frequent checks on the functioning of sensitive areas.

Role of the Biology Teacher in the Laboratory

The biology teacher should be able to ensure that experiments are successfully conducted and the records and reports are well done and prepared by the students. The teacher should among other things:

- Prepare worksheets;
- Summarise experiments for students;
- Check students' records and laboratory reports;
- Give requirements for practical; and
- Order for chemicals, apparatus and equipment.

The School Biology Laboratory Management

The management of the laboratory is the responsibility of the biology teacher or the head of the department of biology. Whereas the technician is important in this task, the responsibility must not be heaped on him or her alone. Laboratory management is mainly concerned with:

- Providing materials, chemicals, and equipment for laboratory work;
- Maintaining a proper inventory of equipment to keep track of the number and type of equipment available in the laboratory;
- Maintenance and repair of equipment, apparatus and fixtures such as taps, sinks, drains, and burners and ensuring that any breakages or blockages of sinks are appropriately fixed. Marinating and repair saves future costs of purchasing the same;
- Ensuring that expensive and fragile equipment and apparatus like microscopes are safely kept in lockable areas;
- Safety in the laboratory; and

- Provision of a laboratory manual to help the teacher and the technician manage the laboratory equipment, materials, apparatus and chemicals in a cost-effective manner.

Safety in the Biology Laboratory

There are many risks in the biology laboratory and it's the teacher's responsibility to prevent any harm to the students. The risks arise from chemicals, fires, explosions, broken glasses, poisonous gas emissions, wastewater and solutions among others. The biology teacher together the laboratory technician should ensure that the laboratory has the required safety apparatus and equipment such as First Aid kit, fire extinguisher, and a set of instructions on safety precautions as well as laboratory rules and regulations that students should be aware of. These should be displayed on the laboratory notice board where the students are able to read them.

First Aid

In the biology laboratory, the common accidents include skin burns from corrosive acids and alkalis and Bunsen burner fires, cuts from broken glassware among others. Sometimes, the students may touch carcinogenic chemicals such as benzene and immediate safety measures should be carried out.

First Aid kits should be available in the laboratory, whose contents must be replenished frequently and expired contents disposed of. The basic contents of the kits include antiseptic cream; tweezers; scissor; bandages; surgical gauze; safety pins; eye lotions; Vaseline; antiseptic fluid; cotton wool; Elastoplast; surgical adhesive tape; 2% acetic acid for alkali burns; 2% sodium carbonate for acid burns; and paracetamol tablets.

Common Injuries in the Biology Laboratory

The biology laboratory is likely to generate fires in one way or another, either from an explosion or from gas leaks. The laboratory should be equipped with a fire extinguisher that is tested regularly. The use of sand baths is also encouraged. The commonest injuries in the biology laboratory include, but not limited, to the following:

- Burns and scalds;
- Cuts and scratches;
- Poisoning; and
- Eye injuries.

Burns and Scalds

Burns arise from contact with dry heat either directly from flames or indirectly from hot surfaces. Scalds are caused by contact with hot liquids. The two injuries are treated by immersing the burnt area in cold water or wrapping it lightly with a wet cloth. Alternatively, gauze may be applied over the burnt area.

Acid and alkali burns from concentrated acids or alkalis are treated by either placing the burnt area under a running tap of water to wash off the acid, by immersing it in a bucket of water. Burns from acids can also be treated by spreading over it some saturated sodium bicarbonate solu-tion and then covering over it with a gauze. Burns from alkalis can also be treated by spreading 1% acetic acid over the area and covering it with a gauze.

Cuts and Scratches

Cuts are commonly caused by carelessly handling broken glassware and sharp blades and knives. Whenever cuts occur, it will be necessary to wash the wound with clean water and then bathing it in antiseptic solution or salt water. If the cut is small, it will be necessary to cover it with an Elastoplast; if it's large, apply antiseptic cream and then cover with gauze and the bandage.

Poisoning

While poisoning may not be common in schools, sometimes the students may touch and later swallow poisonous chemicals such as cyanide, lead, barium, mercury, acids, alkalis, and even carcinogenic (cancer-causing) chemicals such as asbestos and benzene. When this happens, the students are likely to be poisoned. In this case, the teacher is advised to immediately refer the student for medical help. The teacher is expected to advise the students to strictly adhere to the laboratory rules and regulations.

Eye injuries

Common eye injuries in the laboratory are caused by exposure to fumes or vapour from some chemicals and from splashes of corrosive liquids into the eye. Sometimes, specks of grit or even sharp particles like glass may get into the eye. The first aid for the first two cases includes washing the student's eye with a gentle stream of water from a wash bottle. If the grit doesn't get dislodged they may be removed by wiping gently with a clean piece of cloth. For the third case, the student should be taken immediately to the hospital for medical attention.

School Biology Laboratory Rules and Regulations

School laboratories are expected to develop the rules and regulations whose function is to reduce harmful incidents in the laboratories. Each school should have a set of such rules displayed in the laboratory noticeboards. The Ministry of Education has developed prototype rules and regulations and schools are expected to domesticate them for their students and teachers. Some of the rules include:

- No student may enter the laboratory unless prior permission has been granted by the teacher;
- Apparatus and materials in the laboratory must be handled only on instructions from the teacher;
- No apparatus or material may be removed from the laboratory;
- Don't put any chemicals in your mouth. If you do accidentally, spit out at once and rinse the mouth with a lot of water and report the incident to the teacher immediately;
- Don't eat or taste anything while in the laboratory;
- Any injury must be reported to the teacher immediately;
- Any acids or alkalis spilt on the skin or clothes must be washed off at once with much water;
- Don't dispose of solids such as broken glasses in sinks—use waste bins; and
- Bottles must not be held or carried by the neck.

Box 13.1

Discussion Questions

1) Write down as many rules as possible your students should observe in your biology laboratory.
2) Examine your biology laboratory and answer the following questions:
   (a) What is the orientation of the laboratory? Why?
   (b) How are the benches/tables arranged? Why?
   (c) In what direction do the doors and windows open relative to the main laboratory? Why?
   (d) Is the roof fitted with a ceiling?
3) Suppose your two biology classes for the same grade have 100 students, 50 for each class. Your biology laboratory only accommodates 25 students. Draw out a weekly rotation timetable to enable you to conduct the practicals effectively.
4) What do you think would happen to science if all the laboratories in the whole world were to be closed?

Conclusion

The chapter implies that biology being a science, students should be equipped with the skills and attitudes of doing biology the way scientists do science. Therefore, schools are expected to provide these facilities for the purpose of teaching biology. It is only by experience in the school laboratory that students would appreciate how the knowledge of science has come into existence and why the discipline has shaped human history. This chapter has provided information on the rationale for providing the biology laboratory in schools and the minimum equipment, apparatus and chemicals required to promote the learning of the subject in a meaningful manner. The chapter also gives useful information on safety measures, regulations and provision of materials such as realia and preservation of biological specimens.

References

Twoli, N. W (2006) *Teaching Secondary School Chemistry: A Textbook for Teachers in Developing Countries.* Nairobi: Nehema Publishers

Monk, M and Osborne, J. (2002).*Good practice in Science Teaching.* Buckingham: Open University

ASSESSMENT IS often equated and confused with evaluation, but the two concepts are different.

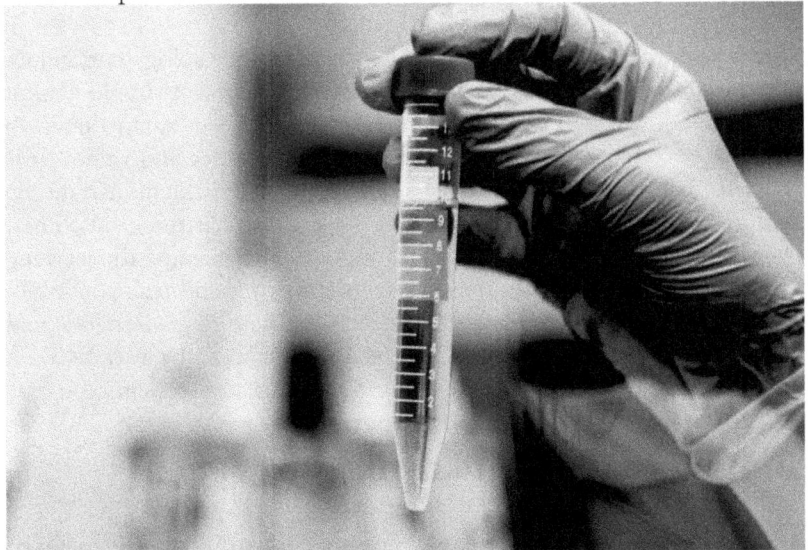

Photo: KONRAD/FLICKR

14

CHAPTER FOURTEEN

EVALUATION OF LEARNING IN BIOLOGY

◆ Definition of Evaluation and Assessment ◆ Types of Evaluation ◆ Purpose of Assessment in Biology ◆ Assessment Techniques in Biology ◆ Preparing a Classroom Test in Biology ◆ Assessing Practical Skills

Introduction

Consistent with the goals, objectives and content of the biology curriculum outlined earlier in this book, students' learning in the subject should consist of cognitive and affective components that enable the students to develop the requisite knowledge and understanding, skills, attitudes and values. It is necessary to determine the extent the students are able to demonstrate the acquisition of these attributes as stated in the expected learning outcomes. This is possible through evaluation which involves determining the learning achievement of pupils through questions, written work and tests and practising self-reflection (the ability to honestly evaluate the strengths and weaknesses of your own teaching). This chapter outlines the various ways of finding out if the students have learned the various attributes in biology.

Definition of Evaluation and Assessment

Evaluation is a decision making process about the effectiveness of teaching and learning. It is the process of determining the value or worth of a programme, course, or another initiative, toward the ultimate goal of making decisions about adopting, rejecting, or revising the innovation. In terms of teaching, it is the process by which information about the students behaviour in respect to what they have learned is collected, quantified and ordered for the purpose of making a decision about the extent of achievement or attainment of outcomes or standards expected of them after attending required lengths of the session in a class.

Proper evaluation in biology involves a wider array of measurements including scores on tests, laboratory reports, field trips, term papers, research projects and other records. Such informal ways of collecting information for comparison purpose constitute **educational assessment. Evaluation in biology, therefore, refers to the assessment of student's progress in the learning of biology.**

Assessment is often equated and confused with evaluation, but the two concepts are different. Assessment encompasses methods for measuring or testing performance on a set of competencies. Assessment is used to determine what a student knows or can do, while evaluation is used to determine the worth or value of a course or programme. Assessment data effects student advancement, placement, and grades, as well as decisions about instructional strategies and curriculum (Herman and Knuth, 1991). Evaluation is the more inclusive term, often making use of assessment data in addition to many other data sources to make decisions about revising, adopting, or rejecting a course or programme.

The device which is used to reveal or measure student behaviour is a **test**. The test itself does not decide who has passed or failed; who has attained a satisfactory standard or below standard. The test maker decides the cut-off point in the testing continuum representing a satisfactory standard or level of achievement.

Types of Evaluation

Evaluation may be formative or summative:

Formative evaluation: This takes place during the course and used for internal feedback to monitor students' progress following instruction and determine their learning difficulties. On the basis of this feedback, improvement of the teachers' strategies may be attained. Classroom tests are examples of formative evaluation.

Summative evaluation: This takes place at the end of the course to establish the students' terminal achievement and is usually conducted by external agencies such as the Kenya National Examinations Council (KNEC). The information resulting from the evaluation is usually used for selection for further education, training and job placement. The KCSE examinations are examples of summative evaluation.

Purpose of Assessment in Biology

- Diagnosing students' weaknesses and strengths in order to plan for remediation;
- Determining students' mastery of content (cognitive attainment), acquisition of scientific skills (both cognitive and psychomotor attainment), and development of scientific attitudes (affective attainment);
- Giving feedback to students on their level of achievement and what they can do and what they need to know and do in order to improve their performance (i.e. setting their operational targets for themselves);
- Helping learners to identify difficult areas which require more time to study;
- Helping teachers to guide students about their future career orientation;
- Informing parents, guardians and sponsors about students' progress;

- Appraising the teacher's own effectiveness of instruction, in terms of achievement of objectives. If the results are poor, this may indicate weakness in the teacher's instructions and may compel them to redirect the efforts and strategies. This may also entail revising schemes of work, lesson planning, and use of resources that may be more effective in achieving the learning objectives;
- Promoting school and national objectives; and
- Providing information in terms of students' capabilities and competencies to be used in promotion to new grades, or in selection for further education, training and job placement.

Assessment Techniques in Biology

The most commonly used techniques for assessing achievement in biology include:

- Oral/Clinical interviews;
- Written tests/examinations (paper - and – pencil tests);
- Practical tests; and
- Project work.

Oral/Clinical Interviews

Oral interviews are usually used for higher degree examination of students' original piece of academic work (theses defence) and for job placement. However, clinical interviews may also be used to determine the students' perceptions, attitudes, and knowledge of specific concepts, issues, and ideas along the approach first used by Jean Piaget to determine children's stages of mental development. Much of the research that has yielded immense information on children's misconceptions or alternative conceptions in science has been achieved through the use of clinical interviews. The approach has several advantages and disadvantages.

Advantages:

- Can be used to asses all cognitive abilities and skills;
- Allow for probing into students' understanding of the issues in question thus revealing any misconceptions they may have;
- Easy to interpret the results of an oral test for diagnostic purposes.

Disadvantages:

- Examiner may be influenced by extraneous factors such as students' verbal fluency and self-confidence;
- Interactive nature of questions may lead to changes in the questions being asked from candidate to candidate thus producing non-comparable results;
- Difficulty in making a judgment after the candidate has left unless the expensive audio-visual resources are used; and
- Difficulty in examining many candidates, making it expensive.

Written Tests/Examinations

These include Essay questions, Objective type questions, and structured questions.

Essay Questions (open-ended)

Essay questions require the candidate to supply more than a sentence to answer the question. The questions thus allow free response and require creativity.

Example 1:

a) What do you understand by the term, gene? (3 marks)
b) Explain how genes function to produce characteristics in organisms (7 marks)

Example 2:

a) When a person sees a large black snake the person suddenly jumps up and screams aloud. With the aid of a large labelled diagram, explain what causes these responses. (10 marks)

The two questions test a wide range of cognitive levels: recall/knowledge, comprehension, application and synthesis. The open-ended questions have various advantages as well as disadvantages.

Advantages:

The advantages of essay type questions are as follows:

- They can be used for assessment of all cognitive abilities;
- They can be used to assess students' communication skills in terms of self-expression and organisation of ideas; and
- They are fairly easy to construct.

Disadvantages:

The disadvantages of essay questions include:

- They take a long time to answer, hence only a few can be attempted by the candidates;
- They usually don't assess the whole syllabus since a few are included in one test/examination;
- They have a high degree of subjectivity in marking and scoring, therefore yielding unreliable results; and
- Students' performance could be influenced by the choice of items.

Objective Questions (closed-ended)

Objective type questions include, but are not limited to, the following categories: Multiple choices, True-false items, matching items, and Completion items. The questions share the following advantages and disadvantages:

Advantages:

- Allow for testing a wide range of cognitive abilities;
- Allow for testing a wide section of the syllabus;
- Allow for analysis and interpretation of responses quite easily and accurately;
- Marking is objective, accurate and easy;
- High validity and reliability of the test; and
- Administration is easy.

Disadvantages:

- It is difficult to construct the items at all cognitive abilities;
- The correct answer could be arrived at by guess work or from the faulty options; and
- They are not suitable for testing students' communication skills.

Examples:

(a) Multiple Choice Questions

Plants lose water through leaves by the process called

    A. Osmosis
    B. Photosynthesis   } Detractors
    C. Evaporation
    D. Transpiration (key)

} Options

b) Matching Questions

In the blank space next to each word in the first column match it with the correct definition by writing the letter for the definition.

1. Watt   (a) Electric force or pressure
2. Amphere flow   (b) A measure of the amount of electric current
3. Volts   (c) A measure of electric consumption
                   (d) A measure of heat formed in a circuit

c) Completion Questions

Three states in which matter exists are ------------, ------------ and ------------ ------------.

d) True/False Questions

In the spaces provided beside each statement write T if it is true or F if it is false.

------- 1. The liver is an organ of excretion

-------- 2. A dry cell changes chemical energy into electrical energy

Structured/Short Answer Questions

These are sets of items thematically related to one another, and which require brief responses only.

Example 1

1) What is the transport medium of each of the following:

   a. Flowering plants---

   b. Mammals--

2) List the substances transported by the transport media in (a) above in the table below:

Transport media in flowering plants	Transport media in mammals

   a. State the similarity in the way materials are transported in flowering plants and in mammals:----------------------------
   ----

Example 2:

The table below shows the energy requirements for both men and women with regard to their various activities. Study the table and answer the questions that follow.

Activity	Energy expenditure(KJ/Min)	
	Woman	Man
Sleeping	4.2	4.2
Sitting and reading	4.9	5.7
Light work	14.8	17.8

(a) Why is energy used during sleep? -------------------------------

(b) Account for the higher expenditure in males compared to females---
----

(c) Name the nutritional element that is usually required in large amounts in females than males. Give reasons for this difference. ------------
----

(d) State two factors that affect energy requirements in human beings. -
----

Objective questions are very popular with examiners. However, they have various advantages and disadvantages.

Advantages:

- They allow for testing of wide sections of the syllabus;
- They may be used to test up to higher cognitive abilities;
- They allow students to formulate responses in a brief manner;
- Marking could be objective and accurate;
- They cannot be answered by guessing;
- Interpretation of results is easy; and
- They are easier to construct than multiple choice items.

Disadvantages:

- They may not be used to test higher cognitive abilities unless a lot of thought goes into constructing them;
- Communication skills not adequately assessed;
- Large-scale administration and marking is often difficult; and
- Some subjectivity by the examiner may occur in deciding the correct response.

Preparing a Classroom Test in Biology

The Steps

The design of test items should be based on the students' expected learning outcomes. The first step in designing a quality assessment is, therefore, outlining goals and objectives or learning outcomes to be tested. This is done by examining the syllabus, and the schemes of work, a record of work done and lesson plans used.

The second step is to divide the content into topics and sub-topics and also the scientific skills to be tested, then determining the type of learning those outcomes represent in terms of the levels of cognitive categories such as memorising/recall of concepts, comprehension/understanding, application of information/concepts to new situations, analysing data, synthesising resources or information to make meaning and evaluation of information or procedures or solutions as well as the various categories of scientific skills (such as observation, classification, communication, measuring, prediction, inferring, hypothesising, interpretation of data, experimentation among others).

The third step involves making a **specification table** or **grid** (Table 14.1). This is a two-dimensional table indicating each cognitive level or skill to be tested listed along one dimension and the content areas of focus specified along with the second axis. The intersection of the two results into composite cells. In each of the cells, the number of questions for assessing each cognitive level and content is indicated. The grid helps to:

- Balance the distribution of the questions in terms of the content so that adequate coverage of the syllabus is ensured;
- Ensure that a wide range of cognitive abilities is tested. It is recommended that the test contains at least 50% testing recall/knowledge, comprehension and application;
- Ensure that not just content, but an adequate level of process skills are tested; and
- Improve the validity of the test.

Table 14.1: The Specification Grid for Assessment of Learning in Biology

Cognitive Level	Content					
	Classification I	Classification II	Cell Structure	Cell Physiology	Nutrition	Total
Recall	2	1	2	0	3	8
Comprehension	1	1	0	2	2	6
Application	1	1	0	2	1	5
Analysis	0	0	2	1	1	4
Synthesis	0	2	1	2	3	8
Evaluation	0	0	2	1	1	4
Total	4	5	7	8	11	30

The fourth step is to consider the types of test items available and how to use them, including objective, structured and open-ended items. It is necessary to use as many types of questions as possible to enable you to cover the syllabus and assess many aspects. It is recommended that the distribution of the various types of questions or items be as follows:

- Objective test   20%
- Essays   20%
- Structured items   35%

- Practical work   15%
- Project   10%

The fifth step is to construct the questions, putting marks to each part of the questions. It helps a lot for the teacher to draft as many questions as possible on each topic during lesson planning and implementation. The marks allocated to each question should reflect the level of difficulty of the question. The questions should be arranged in order of difficulty with the hardest ones at the end of the test and the reading level of the questions should be kept as low as possible. The instructions should be clear, complete and precise. The use of ambiguous words should be avoided.

The sixth step involves preparation of the marking scheme before marking the scripts. It is advantageous for the teacher to make the marking scheme at the time of drafting the items. After marking it is necessary to analyse the results to help the teacher in improving the instructional strategies and in identifying the learning difficulties pupils experienced.

Preparing Marking Scheme

In making a marking scheme for open-ended questions, marking points should be as flexible as possible to allow for alternative points of view. Allocate the marks for specific points, denoted by a semicolon.

Example 1

a) What do you understand by the term gene?   (3 marks)
b) Describe how genes function to produce characteristics in organisms.   (7 marks)

Marking Scheme

a) The functional unit of DNA/nucleotide sequence; or other words to that effect (OWTTE). Consists of phosphate, sugar, and nitrogen base; Genes are located on chromosomes; (3 marks)
b) Specific genes are located at specific loci on chromosomes; each gene is composed of specific lengths of DNA; that determine the characteristics of an organism. Through meiosis and reproduction, genes are transmitted from one generation to another; allowing characteristics to be passed on as well. The phenotypes displayed

by an organism is the result of the ability of the genes to express themselves; (7 marks)

Assessing Practical Skills

Biology consists of conceptual and procedural knowledge. The former deals with theoretical understandings while the latter with the scientific process of acquiring biological knowledge. The assessment techniques discussed earlier mainly focused on theoretical forms of knowledge. However, the teacher also needs to assess the students' acquisition of scientific or practical skills through the learning of biology. The students are expected to demonstrate that they have reached an acceptable standard of ability to correlate theory with the necessary practical skills in biology.

The biology curriculum requires schools to assess practical skills during routine laboratory practical lessons and at the end of the school term for the entire time at the secondary school level. In addition, the students are assessed at the end of the secondary school course through an external examination (the practical paper).

The practical paper may involve any of the following:

- Manipulative or handling(psychomotor) skills; and
- Science process skills.

Assessment of Manipulative Skills

The manipulative skills assessment involves an examination of the ability to handle, arrange, pour, cut, wash, and dissect, among other skills. The practical skills tested include the following:

- Section cutting and simple staining of plant parts, and preparation of temporary slides for microscopic examination of these preparations;
- Interpreting what is observed from permanent slides under the microscope;
- Dissection of part of an animal or plant organ and the interpretation of the anatomy displayed;
- Drawing accurate and well-labelled diagrams of plant and animal specimens either whole or dissected or sectioned;
- Classification of plants and animals using observable features or the dichotomous keys;

- Identification of whole or parts of plants or animals based on specific reasons;
- Identification of plant and animal specimens based on the morphological and anatomical features and adaptation to their habitats; and
- Manipulation of specimens, reagents, and apparatus involving simple physiological phenomena such as osmosis, plasmolysis, enzyme action, food tests, permeability, turgidity among others.

The success of these assessments will depend on the provision of proper materials or specimens, apparatus, and accurate observations and records of what is observed or done.

Example:

You are provided with the leaves of the following plants:

a) Jacaranda
b) Hibiscus
c) Maize
d) Nandi flame
e) Bougainvillaea

Examine the leaves carefully and construct the likely dichotomous key for their identification.

Box 14.1

Discussion Questions

1) In constructing a dichotomous key observable morphological features are used as much as possible. Which features were employed in constructing the dichotomous key of the leaves?
2) Provide a logical flow of the following guidelines for the construction of the above dichotomous key:
a) Generalisations and overlapping characteristics such as height should be avoided as much as possible
b) Major characteristics such as leaf type and veneration are used first It is usual to list the key features of each organism which are then used while constructing the dichotomous key

c) Minor characteristics such as the presence or absence of a terminal leaflet should be used later in the process
d) A dichotomous key is constructed by separating or branching of organisms or their parts into two at a time according to their morphological features based on evolutionary relationships
e) Paired statements that describe the variation in the characteristics are based on the presence of the features and absence of the features

Assessment of Science Process Skills

The science process skills (also referred to as scientific skills) are more of cognitive skills and involve the process of scientific investigation. The skills include observation; raising questions; hypothesising; planning experiments; controlling variables; measuring; experimenting; quantifying; recording; estimating; communicating; comparing; predicting; inferring; interpretation among others.

The practical test is expected to assess many of these cognitive skills together with manipulative skills in an integrated manner. The test items on manipulation of plant and animal specimens focus on manipulative skills while items on physiological processes such as food test, osmosis, plasmolysis, and turgor pressure usually focus on acquisition and use of science process skills. The major weakness of items on cognitive skills is that it is difficult to ascertain if the records put on paper by the individu-al is what is actually observed. The supervisor or teacher must surely ensure that the students record what they have observed.

Example 1:

You are provided with substances in solution labelled A, B and C, and 10% sodium hydroxide solution. You are also provided with test tubes, droppers, and a 10ml measuring cylinder. Conduct the following test:

a) Put $2cm^3$ of solution A into a clean test-tube. Add an equal amount of 10% sodium hydroxide solution and shake well.
b) Into the mixture, add solution B drop by drop shaking after every addition.
c) Repeat the procedure for solution C.

d) Record your observations in the table below.
e) Provide the identity of solutions A, B and C.

Table 14.2: Observation and Inference table

Test	Observation	Inference
1.		
2.		
3.		
5.		
6.		
Identity the likely solutions represented by the letters A, B, and C		

Box 14.2

Discussion

Identify the manipulative scientific skills and the scientific skills that student is expected to employ in this practical task.

Conclusion

This chapter has bluntly implied that it would not make sense if teachers engage students in the learning process and they don't bother to find out if the students have actually learned or not. Students' learning in biology should consist of cognitive and affective components that enable the students to develop the requisite knowledge, skills, attitudes and values. The biology teachers are expected to determine the extent the students are able to demonstrate the acquisition of these attributes as stated in the instructional objectives. Evaluation and assessment procedures which involve determining the achievement of students in biology through questions, written work and tests, practical performance and practising self-reflection have been suggested.

References

Herman and Knuth, (1991). *What does research say about assessment?* (Downloaded on 21/05/2010 from http://methodenpool.uni-koelnide/portifolio/what does research say about assessment.htm)

KNEC REPORT have over the years identified some of the concepts in biology that students have usually found it difficult to understand

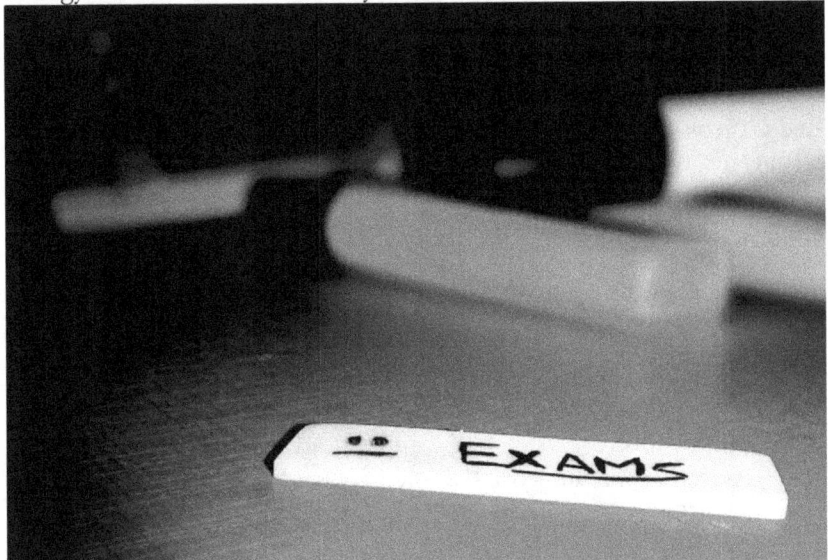

Photo: LUSTROUS SMILE/FLICKR

15

CHAPTER FIFTEEN

HINTS ON TEACHING SOME DIFFICULT CONCEPTS IN BIOLOGY CURRICULUM

- ◆ Some Difficult Biology Concept in the Curriculum in Kenya
- ◆ Highlights of Key Teaching Approaches of the Difficult Biology Concepts

Introduction

The Kenya National Examinations Council (KNEC) reports have over the years identified some of the concepts in biology that students have usually found difficult to understand. The reports have often identified the following concepts as being problematic to the students whenever they sit for the summative evaluation examinations at the end of their secondary school education: The concepts fall under the following topics:

- Cell physiology;
- Plant and animal nutrition;
- Transport in plants and animals;
- Gaseous exchange in plants and animals;
- Respiration;
- Excretion and homeostasis;
- Classification;
- Ecology;
- Genetics;
- Evolution; and
- Reception, Response and Coordination.

This chapter identifies the key concepts that students find difficult to learn across all the four years of study and the key suggestions are made regarding the approaches and resources required to learn the concepts with ease.

Some Difficult Concepts in the Biology Curriculum in Kenya

The following concepts are among the most difficult for learners of biology in Kenya:

Form 1

- The relationship between Surface area and volume;
- Osmotic pressure;
- Active transport;
- Light and dark stages of photosynthesis;

- Carbohydrates: Condensation, hydrolysis, reducing and non-reducing sugars;
- Tests for reducing and non-reducing sugars;
- Colloidal suspensions in proteins;
- Enzyme co-factors and co-enzymes;
- Emulsification of fats; and
- Basal Metabolic Rate and energy requirements.

Form 2

- Forces in the transportation of water and mineral salts in plants: transportation pull, adhesion and cohesion, capillarity, root pressure;
- Mechanism of opening and closing stomata;
- Mechanism of gaseous exchange in gills of bony fish;
- Glycolysis and Krebs cycle; aerobic and anaerobic respiration;
- Oxygen debt; and
- Filtration in kidneys.

Form 3

- Binomial nomenclature in classification;
- Construction and use of dichotomous keys;
- Estimation of the population;
- Mitosis and meiosis; and
- Dormancy in seeds.

Form 4

- Continuous and discontinuous variation;
- Incomplete dominance;
- DNA structure and replication;
- Linkage;
- Crossing over;
- Evidence for organic evolution;
- Mechanism of evolution;
- Transmission of an impulse in muscle cells; and
- Negative feedback mechanism for thyroxin.

Highlights of the Key Teaching Approaches of the Difficult Biology Concepts

The following suggestions in Table 15.1 may help the biology teacher to plan and prepare for teaching some of the concepts perceived to be difficult by students.

Table 15.1: Suggested Teaching Approaches for Selected Concepts in Biology

Form	Topic	Concept	Teaching Approach and Activities	Teaching Resources
1	Cell physiology	Surface Area and Volume	-Whole class discussion on the significance of diffusion in the absorption of important materials into the cell and the role of surface area and surface area to volume ratio in this. -Group work to calculate surface area and volume of cuboids and the resultant SA/V ratio. -Group work to measure and calculate the SA/V ratio of various 3-D objects and present results graphically. -Whole class discussion on applications of SA/V	-Secondary Biology Students' BK 1, pp37-43 -Diagrams and pictures on Charts -Relevant data
		Osmotic Pressure	-Whole class discussion on meaning and significance of osmotic pressure and its relation to osmotic potential, concentration gradient, and turgor pressure. -Group experiments on	-Secondary Biology Students' BK 1, pp45-49 -Worksheet or instructions on

			osmosis in plant cells focusing on plasmolysis, deplasmolysis, wall pressure, turgor pressure and flaccidity	pp44-48
				-Cork borer 0.5 diameter, scalpel, Irish potato, rulers, boiling tubes, distilled water, strong sugar solution, tissue paper, blotting paper, microscope slides, droppers, forceps, forceps, onion epidermis, light microscope, 5 % salt solution, newspaper prints.
			-Demonstration experiment on haemolysis and crenation in animal cells, using mammalian red blood cells obtained from the butcher	
			-Ensuring the experiment reports are written with a clear interpretation of data.	
		Active transport	-Whole class discussion of meaning and significance of active transport	-Secondary Biology Students' BK 1, pp45-49
			-Whole class discussion using examples to provide application of active transport in plant and animal cells	
	Nutrition in Plants and Animals	Process of Photosynthesis	-Whole class discussion focusing on the meaning, significance and mechanism of the light-dependent stage or photolysis and its products.	-Secondary Biology Students' BK 1, pp57-58
			-Whole class discussion	

				focusing on the meaning, significance and mechanism of light independent or dark stage or carbon IV oxide fixation stage and its products	
			Carbohydrates	-Whole class discussion focusing on the chemical composition of carbohydrates that constitute plant and animal living matter. -Experiments focusing on food tests that identify, using Benedict's solution, monosaccharide(reducing sugars), disaccharide(non-reducing sugars) and polysaccharide(non-sugars) -Group work to explain the processes of condensation and hydrolysis in carbohydrate functioning in living organisms.	-Secondary Biology Students' BK 1, pp63-65 -Worksheet or instructions in the textbook -Test-tubes, Benedict's solution, dil. hydrochloric acid, sodium hydrogen carbonate solution, hot water bath, droppers, 10 ml measuring cylinders, Bunsen burners., food substances in solution form.
2		Transport in Plants and Animals	Capillarity, Transpiration Pull, Cohesion and Adhesion and Root Pressure Forces involved in Transporta-	-Demonstration experiments to illustrate capillary action and the rate of transpiration. -Whole class discussion focusing on the role of xylem structure in influencing transpiration	-Secondary Biology Students' BK 2, pp11-17 - Secondary Biology Teachers' Guide BK 2,

		tion of water and mineral salts	pull, and capillary action, cohesion and adhesion and root pressure.	pp4-5 -Charts illustrating transpiration pull, capillary action, root pressure -Glass tubing of different diameter, capillary tube, cobalt chloride paper, desiccator bottle, beaker, water with dye, photometer, leafy shoot, exudation from cut stumps or branches of trees.
	Gaseous Exchange	Gaseous exchange in plants: Opening and closing of stomata	-Whole class discussion focusing on three theories of stomata opening and closing noting the suitability of each. -Demonstration of opening and closing of stomata by examining leaf epidermis mounted in a drop of water and in a drop of salt solution respectively on the slide and viewed under a microscope. Guide the students to note that the structure of the guard cells holds the	-Secondary Biology Students' BK 2, pp58-60 - Secondary Biology Teachers' Guide BK 2, pp28-29 -Charts illustrating changes in guard cells in the closing and opening of stomata.

				key to the mechanism.	-Leaf epidermis, glass slides, water, droppers, microscope, salt solution
			Gaseous Exchange in Animals: Mechanism of gaseous exchange in gills of bony fish	-Whole class discussion focusing on the role of the structure of gills with respect to blood capillaries in influencing gaseous exchange in fish through the countercurrent flow system model -Demonstration using Liebig's condenser as an analogy to explain countercurrent flow system -Observation of live fish to identify breathing movements in the fish	Secondary Biology Students' BK 2, pp66-68 - Secondary Biology Teachers' Guide BK 2, pp31-32 -Charts illustrating the countercurrent flow system across the gills. -Bony fish, scalpel, water in a beaker, hand lens, Basin with water, dissecting microscope
			Respiration: Glycolysis, Kreb's cycle in aerobic and anaerobic respiration and the oxygen debt.	-Whole class discussion focusing on glycolysis and Kreb's cycle and how the processes differ in aerobic and anaerobic respiration, including fermentation and oxygen debt. -Group experiments to determine gas emitted when food is burnt,	-Secondary Biology Students' BK 2, pp85-93 - Secondary Biology Teachers' Guide BK 2, pp42-43 -Charts illustrating the

			production of heat in germinating bean seeds, and gas produced during fermentation	structure of mitochondrion, glycolysis and Kreb's cycle processes, aerobic and anaerobic and equations summarizing the processes.
				-Foodstuffs in powder or crushed form, boiling tubes, test tubes, calcium hydroxide solution, rubber stoppers, anhydrous cobalt chloride paper, Bunsen burner, delivery tubes, retort stands, boiled bean seeds, soaked bean seeds, vacuum flasks, thermometers, methanol, measuring cylinders, 10% glucose solutions, yeast, oil,
	Excretion and Homeo-	Filtration in Glomerulus of Human Kid-	-Whole class discussion focusing on how the structure of the human	-Secondary Biology Students' BK 2,

| | | stasis | ney | kidney is suited to its functions including glomerulus, nephron, Bowman's capsule, Malpighian body, and a loop of Henle

-Group discussion on the mechanism of filtration in the kidney focusing on the counter-current mechanism in the loop of Henle and the associated terms like

Ultrafiltration and selective reabsorption and the role of hormones.

-Demonstration experiment on the composition of mammalian urine such as reducing sugars, proteins, chlorides, minerals, and pH | pp106-110

- Secondary Biology Teachers' Guide BK 2, pp49

- Charts illustrating the structure of nephron and the possible explanation of the filtration mechanism |
|---|---|---|---|---|---|
| 3 | Classification | Binomial nomenclature and dichotomous keys in classification | -Whole class discussion of meaning and significance of binomial nomenclature and the fundamental features used in the classification

-Demonstration of construction and use of dichotomous key including a summary of the schematic framework for the construction.

-Group work on construction and use of | -Secondary Biology Students' BK 3, pp1-11

- Secondary Biology Teachers' Guide BK 2, pp1-3

- Charts illustrating the procedures used, and the key features considered, in |

				dichotomous keys to classify living things	the construction of dichotomous keys. -Various types of plants and animal specimens provided for construction of the keys
			Estimation of Population	-Whole class discussion focusing on methods of the population estimate, namely sampling and total count -Demonstration on estimating population using total count method and sampling method(quadrat, transect, capture-recapture, percentage cover, and frequency) - Simulation studies to estimate the population. -Project work to conduct ecological studies including estimation of population in the community around the school	-Secondary Biology Students' BK 3, pp1-11 - Secondary Biology Teachers' Guide BK 2, pp1-3 - Charts illustrating the procedures used in population estimate. -The school /community environment and tools for ecological studies such as quadrants, tape measures, etc
			Mitosis and Meiosis	-Whole class discussion focusing on the significance of chromosomes in reproduction and the	- Secondary Biology Students' BK 3,

				role of mitosis and meiosis	pp60-69
				-Demonstration of chromosomes and the changes in them during mitosis and meiosis focusing on diploid and haploid state	- Secondary Biology Teachers' Guide BK 3, pp20-25
				-Leading students to analyses the stages in mitosis and meiosis and to differentiate between them.	- Charts illustrating cell division and behaviour of chromosomes
				-Groups to examine and draw stages of mitosis using young onion tips, charts and electron micrographs	- Chromosomes represented by beads threaded through a wire while; genes represented by the beads on the wire; Plasticine or moulding clay used to show behaviour of chromosomes during division.
				-Groups examine and draw stages of meiosis using anthers of a flower	
4		Genetics	Continuous and discontinuous variation	-Whole class discussion on meaning, sources and significance of variation	- Secondary Biology Students' BK 4, pp1-10
				-Conduct group investigations on chromosome behaviour in mitosis and meiosis by	- Secondary Biology Teachers' Guide BK 4,

			using plasticine or moulding clay, focusing on crossing over and independent assortment during gamete formation. -Group investigations on continuous and discontinuous variation focusing on fingerprints and tongue rolling, the height of students in class and length of eucalyptus leaves	pp1-4 - Charts illustrating cell division and behaviour of chromosomes during gamete formation -Thick coloured thread, scissors, white manila paper, transparent cellotape, felt pens of different colours, Worksheet or instructions in the textbook	
		Genes and DNA	-Whole class discussion focusing on the structure of DNA and how it replicates itself as a mechanism important in heredity. -Demonstrate the role of DNA in protein synthesis and the ultimate control of inherited characteristics	- Secondary Biology Students' BK 4, pp10-13 - Secondary Biology Teachers' Guide BK 4, pp1-4 -Charts illustrating DNA replication and protein synthesis. -Coloured illustration of	

				DNA downloaded from websites
		Mendel's First Law of Heredity: Complete and incomplete dominance	-Demonstration focusing on characteristics exhibiting complete and incomplete dominance using as many examples as possible -Group work involving throwing a coin to show the probability of fusion of gametes and random fusion of gametes using different coloured beads	- Secondary Biology Students' BK 4, pp13-27 - Secondary Biology Teachers' Guide BK 4, pp1-8 -Charts illustrating complete and incomplete dominance
		Linkage	-Whole class discussion focusing on meaning, sources and significance of linkage or linked genes. -Demonstration of sex-linked characteristics such as haemophilia, colour blindness, hairy ears and effect of crossing over on linked characteristics	- Secondary Biology Students' BK 4, pp27-32 - Secondary Biology Teachers' Guide BK 4, pp1-13 -Charts illustrating linkage and sex linkage
	Evolution	**Origin of life**	-Class discussion focusing on meaning and significance of evolution -Class discussion on the origin of life as explained by the theory of organic evolution and also theory of spe-	- Secondary Biology Students' BK 4, pp53-55 - Secondary Biology Teachers' Guide BK 4,

				cial creation	pp20
					-Charts illustrating the origin of life
		Evidence of organic evolution	-Whole class discussion focusing on the key evidence of organic evolution such as fossil record, geographical distribution, comparative embryology, and comparative anatomy	- Secondary Biology Students' BK 4, pp 55-70
					- Secondary Biology Teachers' Guide BK 4, pp20-22
					-Charts illustrating evidence of evolution such as embryonic development, forelimbs of different vertebrates, wings of birds and insects, pericarp modifications, and cell organelles
					-world map showing how various geographical regions relate to the fauna and flora
		Mechanism of Evolution	-Class discuss focusing on theories of evolution by Lamarck and	- Secondary Biology Students' BK 4,

			Darwin and their limitations. Indicate the link of genetics to evolution since genetic changes are a major cause of variation

-Demonstration of natural selection in peppered moths (adaptive radiation), and resistance to drugs, pesticides, and antibiotics. Emphasis should be on the role of variation in evolution.

-Group discussion on the possible causes of variation in evolution—such as mutations, pollution, food types, etc | pp 71-76

- Secondary Biology Teachers' Guide BK 4, pp22-23

- Charts, pictures, illustrating role variation in evolution

-Worksheet on the role of variation in evolution---see also revision questions on Students' book pages 77 and Teachers' Guide pages 23-26 |
| | Response and coordination in Plants and Animals | Nervous System in Mammals: Transmission of Nerve Impulse | -Demonstration of the transmission of nerve impulse on a nerve axon with emphasis on the role of sodium and potassium ions. Attention should be placed on the key aspects including resting potential, active potential, sodium pump, Polarization and depolarization, and nerve impulse. | - Secondary Biology Students' BK 4, pp 98-99

- Secondary Biology Teachers' Guide BK 4, pp 37

- Charts showing illustrations of interchange of sodium and potassium |

				ions thus creating the movement of action potential on the axon constituting a nerve impulse
		Transmission of a nerve impulse across the neuro-junction or synapse	Demonstration of the transmission of a nerve impulse across the synapse with the help of acetylcholine which is produced when the nerve reaches the junction. The substance makes membrane permeable allowing sodium ions in and creation of action potential which is transmitted as a nerve impulse. The enzyme cholinesterase then destroys the transmitter substance	- Secondary Biology Students' BK 4, pp 99-100 - Secondary Biology Teachers' Guide BK 4, pp 37 - Charts showing illustrations of impulse transmission across neuro-junction.

Conclusion

This chapter has made it clear that the Kenya National Examinations Council (KNEC) reports have over the years identified some of the concepts in biology that students have usually found it difficult to understand. From the analysis, in this book, theoretical concepts that are multi-ordinal in nature are usually abstract and difficult for most learners to understand. The reports have often identified some concepts which are spread across the curriculum from Form one to Form Four as being problematic to the students whenever they sit for the summative evaluation examinations at the end of their secondary school education: The topics under which these concepts fall are provided in this chapter. The chapter highlights some of the ways in which biology teachers may effectively teach such concepts to facilitate students' effective learning

References

Republic of Kenya (2018) *Secondary Biology Form One Students' Book 5th Edn.* Nairobi: Kenya Literature Bureau Publishers and Printers

Republic of Kenya (2018) *Secondary Biology Form One Teachers' Guide 5th Edn.* Nairobi: Kenya Literature Bureau Publishers and Printers

Republic of Kenya (2018) *Secondary Biology Form Two Students' Book 5th Edn.* Nairobi: Kenya Literature Bureau Publishers and Printers

Republic of Kenya (2018) *Secondary Biology Form Two Teachers' Guide 5th Edn.* Nairobi: Kenya Literature Bureau Publishers and Printers

Republic of Kenya (2018) *Secondary Biology Form Three Students' Book 5th Edn.* Nairobi: Kenya Literature Bureau Publishers and Printers

Republic of Kenya (2018) *Secondary Biology Form Three Teachers' Guide 5th Edn.* Nairobi: Kenya Literature Bureau Publishers and Printers

Republic of Kenya (2018) *Secondary Biology Form Four Students' Book 5th Edn.* Nairobi: Kenya Literature Bureau Publishers and Printers

Republic of Kenya (2018) *Secondary Biology Form Four Teachers' Guide 5th Edn.* Nairobi: Kenya Literature Bureau Publishers and Printers

REVISION QUESTIONS

QUESTION ONE

(a) Science is one of the most established disciplines.
   i. Give any **three** reasons why biology is considered as a science (3 marks)
   ii. Distinguish between biology and biology education (3 marks)
   iii. State and explain the steps in the scientific method (5 marks)
   iv. State any three goals of biology education. (3 marks)
   v. Give **three** examples **each** of scientific theories and scientific laws in biology and state their attributes. (6 Marks)

(b) A Form three student carried out a school project to find out the effect of DAP fertilizer (measured in capfuls or cf) on the growth of <u>Zebrina</u> houseplants. He recorded the results as shown in Table 1 below.

Table 1: Total Vine Length in Centimeters (cm)

ITEM	VINE LENGTH PER FERTILIZER DOSE				
Dose of Fertilizer	0cf	½ cf	1 cf	1½ cf	2 cf
Length of vine (0 weeks)	145	150	148	151	149
Length of vine (6 weeks)	166	180	181	179	184
Increase in vine length	21	30	33	28	35

c) Calculate the percentage increase in vine length in the plant which received fertilizer compared to the one without fertilizer in the cases receiving the following amounts:

   i. One-half (½) capful of fertilizer
   ii. One (1) capful of fertilizer
   iii. One and one-half (1½) capfuls of fertilizer
   iv. Two (2) capfuls of fertilizer (4 marks)

d) State the likely hypothesis the student formed for this project. (2 marks)

e) State any **FOUR** control variables the student identified for the project (2marks)
f) What was the likely conclusion made by the student from the results? (2 Marks)

QUESTION TWO

Effective planning is crucial for the successful teaching of biology.
(a) Discuss any **FIVE** advantages of lesson planning for the purpose of teaching biology (5 marks)
(b) You plan to conduct a two-period biology lesson in Form 4 on the topic: 'Construction and use of simple dichotomous keys in the classification of plants: The case of Malvaceae family'.
  i. State **THREE** knowledge and **THREE** skills objectives you would construct to guide the lesson (6 marks)
  ii. Construct the lesson development component you would use to teach the class (6 marks)
  iii. Make brief lesson notes you would use to conduct the lesson (3 marks)

QUESTION THREE

(a) State any **FIVE** advantages of using teaching aids in the teaching of biology (5 marks)
(b) Discuss the rationale for using teaching aids in biology (5 marks)
(c) Describe how you would use an official class textbook of biology to teach a form 3 class the topic: 'Photosynthesis'. (10 marks)

QUESTION FOUR

(a) State any **FIVE** objectives of the defunct SSP biology course in Kenya (5 marks)
(b) Briefly compare and contrast the SSP biology curriculum with the current secondary school biology curriculum in Kenya (10 marks)
(c) Discuss any **FIVE** factors that led to the failure of the SSP biology course in Kenya (5 marks)

QUESTION FIVE

A biology teacher has developed a test in which the following essay test item was included to assess the Form 4 students' learning of the 'gene concept'.

    (i)   What do you understand by the term 'gene'?
    (ii)   Explain how genes function to bring about the inheritance of characteristics in human beings.

a) Outline any **FOUR** advantages of using a specification grid in planning for the construction of test items in assessing students' learning of biology.
(4 marks)
b) Make a marking scheme for this essay test item.   (10 marks)
c) Outline **THREE** advantages and **THREE** disadvantages of using essay type test items in students' learning of biology   (6 marks)

QUESTION SIX

a) Science is one of the most established disciplines.

  i.   Outline any **FOUR** building blocks of scientific knowledge
                                                                             (4 marks)
  ii.   Give **FOUR** examples of scientific theories in biology and state their attributes   (4 marks)

b) Biology is considered as a science. State **THREE** reasons as to why this is true?   (3 marks)

c) A group of Form 3 students hypothesised that the amount of alcohol produced in fermentation depended on the amount of glucose supplied to the yeast. They planned to use 5%, 10%, 15%, 20%, 25%, and 30% glucose solutions.
  i.   What is the independent variable?   (1 mark)
  ii.   What is the dependent variable?   (1 mark)
  iii. What control treatment should be used?   (2 marks)
  iv. State **ANY TWO** variables which should be controlled for
                                                                             (2 marks)

d) The students used The Scientific Method to carry out their investigation. Describe briefly the **FOUR** major steps in the scientific method   (8 marks)

QUESTION SEVEN

Effective planning for instruction is crucial in the teaching of biology.

    a) Describe briefly **FIVE** things to consider when planning a lesson (5 marks)
    b) Outline the functions of the major components of a lesson plan (10 marks)

QUESTION EIGHT

    a) Distinguish between Assessment and Evaluation in the teaching of biology (4 marks)
    b) Outline **TWO** strengths and **TWO** limitations of using essay type items in testing students' learning in biology (4 marks)
    c) For assessment and evaluation to be effective, a teacher has to go through the process of test construction. Briefly describe **ANY SEVEN STEPS** to effective test construction (7 marks)

QUESTION NINE

    a) Outline the objectives of teaching biology in the secondary schools of Kenya (6 marks)

    b) Outline any **FIVE** problems that militate against the successful realization of the objectives of teaching biology in the secondary schools of Kenya (5 marks)
    c) Discuss any **FOUR** factors that led to the failure of the SSP biology course in Kenya (4 marks)

QUESTION TEN

"What children learn depends not only on what they are taught but also how they are taught, their development level, and their interests and experiences.... These beliefs require that much closer attention is paid to the methods chosen for presenting material..."

    a) Describe **THREE** factors to consider in the selection of teaching methods and techniques when you plan to each a biology concept (6 marks)

b) What **FIVE** reasons would lead you to select 'Demonstration' as one of your teaching techniques? (5 marks)

c) Explain how you would increase the effectiveness of a demonstration lesson (4 marks)

QUESTION ELEVEN

Several changes have taken place in the secondary school biology curriculum in Kenya since independence.

a) Discuss the **FIVE** factors influencing curriculum change in Kenya (5 marks)

b) Discuss the phases that the biology curriculum went through in Kenya since independence (10 marks)

www.ingramcontent.com/pod-product-compliance
Lightning Source LLC
Chambersburg PA
CBHW051203170526
45158CB00013B/85